Occupational Health Law

Occupational Health Law

Second Edition

Diana M. Kloss

*LLB (London), LLM (Tulane), Barrister (Gray's Inn),
Senior Lecturer in Law, University of Manchester*

OXFORD

BLACKWELL SCIENTIFIC PUBLICATIONS

LONDON EDINBURGH BOSTON
MELBOURNE PARIS BERLIN VIENNA

© Diana, M. Kloss 1989, 1994

Blackwell Scientific Publications
Editorial Offices:
Osney Mead, Oxford OX2 0EL
25 John Street, London WC1N 2BL
23 Ainslie Place, Edinburgh EH3 6AJ
238 Main Street, Cambridge,
  Massachusetts 02142, USA
54 University Street, Carlton,
  Victoria 3053, Australia

Other Editorial Offices:
Librairie Arnette SA
1, rue de Lille
75007 Paris
France

Blackwell Wissenschafts-Verlag GmbH
Düsseldorfer Str. 38
D-10707 Berlin
Germany

Blackwell MZV
Feldgasse 13
A-1238 Wien
Austria

First edition published 1989
by BSP Professional Books
Second edition published 1994

Set by DP Photosetting, Aylesbury, Bucks
Printed and bound in Great Britain by
Hartnolls Ltd, Bodmin, Cornwall

DISTRIBUTORS

Marston Book Services Ltd
PO Box 87
Oxford OX2 0DT
(*Orders:* Tel: 0865 791155
      Fax: 0865 791927
      Telex: 837515)

USA
Blackwell Scientific Publications, Inc.
238 Main Street
Cambridge, MA 02142
(*Orders:* Tel: 800 759-6102
           617 876 7000)

Canada
Oxford University Press
70 Wynford Drive
Don Mills
Ontario M3C 1J9
(*Orders:* Tel: (416) 441-2941)

Australia
Blackwell Scientific Publications Pty Ltd
54 University Street
Carlton, Victoria 3053
(*Orders:* Tel: 03 347-5552)

British Library
Cataloguing in Publication Data
A Catalogue record for this book is available
from the British Library

ISBN 0–632–03651–6

Library of Congress
Cataloging in Publication Data
Kloss, Diana M.
    Occupational health law/Diana M.
Kloss. – 2nd ed.
        p.   cm.
    Includes bibliographical references
and index.
    ISBN 0–632–03651–6
    1. Industrial hygiene–Law and
legislation–Great Britain.   2. Industrial
safety–Law and legislation–Great Britain.
I. Title.
KD3168.K58   1994
344.41'0465–dc20
[344.104465]   94-6710
                                   CIP

Contents

Provision and Use of Work Equipment Regulations
(PUWER) 1992 – Personal Protective Equipment
Regulations 1992 – Manual Handling Operations
Regulations 1992 – Health and Safety (Display Screen
Equipment Regulations) 1992 – Protection of the en-
vironment

redundancy – Contravention of a statute – Some other
substantial reason for dismissal – The importance of
procedure – Strikes and industrial action – Time off for
trade union activities

Preface to the Second Edition

Most of my professional life has been spent in the study and practice of employment law and health services law. Occupational health combines these two interests and is an area of growing concern. I have been greatly assisted by members of the medical, nursing and legal professions. The original inspiration came from Professor W.R. Lee, now Emeritus Professor of Occupational Medicine in the University of Manchester, who first introduced me to the discipline. I have benefited also from working with his successor, Professor Nicola Cherry. I am greatly indebted to Marion Balcombe, Jan Rose, Geoff Burrows, Gordon Parker, George Fletcher, Wilf Howe, Elizabeth McCloy, Jim Jolley, Richard Marcus and Tom Whitaker, all of whom pointed out various errors or helped with background material. Any flaws which remain are, of course, my responsibility alone.

My thanks are due to colleagues in the Faculty of Law, especially to Margot Brazier, for whose wise counsel, penetrating analysis and warm encouragement I am sincerely grateful. Hazel Carty and Anthony Ogus made many helpful comments. Richard Miles from Blackwell Scientific Publications assisted with his usual patience and good humour. The law is always changing, but I have tried to incorporate developments up to the end of January 1994.

The book is written from the standpoint of the health professional working in occupational health (OH). Chapter 1 is concerned with legal requirements for the provision of OH services, and with proposals for change. Chapters 2 and 3 deal with the legal status and liability of the OH doctor and nurse, and with the particular problem of confidentiality. OH professionals are regularly concerned with routine medical examinations and health surveillance, discussed in Chapter 4.

The rest of the book seeks to provide an informative background to OH practice. Chapter 5 covers the criminal aspects of health and safety at work and the powers of the inspectorates. Chapters 6 and 7 concern compensation for accidents at work and work-related disease. Chapter 8 surveys the law of employment as it is relevant to the OH practitioner. Finally, Chapter 9 discusses special legal protection for women workers, young workers and those from minority ethnic groups.

Shortly before the first edition of this book was published I learned that Günther, my husband, was suffering from terminal cancer. I am glad that he lived to see in print the product of all his support and encouragement. The second edition has had to be written without him, but is dedicated to his memory.

Diana Kloss
Manchester

Glossary of Abbreviations

The following abbreviations and acronyms have been used in the text:

ACAS:	Advisory, Conciliation and Arbitration Service
ACDP:	Advisory Committee on Dangerous Pathogens
ACGM:	Advisory Committee on Genetic Manipulation
AD:	Appointed Doctor
AFOM:	Associate Member of the Faculty of Occupational Medicine of the Royal College of Physicians
AMA:	Adjudicating Medical Authority
AO:	Adjudication Officer
BMA:	British Medical Association
CEN:	Comité Européen de la Normalisation
CENELEC:	Comité Européen de la Normalisation Electronique
CJEC:	Court of Justice of the European Communities
COSHH:	Control of Substances Hazardous to Health (Regulations and Approved Code of Practice) 1988
DAU:	Data Appraisal Unit
DH:	Department of Health
DSS:	Department of Social Security
EAEC:	European Atomic Energy Community
EC:	European Community
ECSC:	European Coal and Steel Community
ECJ:	European Court of Justice
EMA:	Employment Medical Adviser
EMAS:	Employment Medical Advisory Service
EMSU:	Epidemiology and Medical Statistics Unit
ENA:	Employment Nursing Adviser
EPCA:	Employment Protection (Consolidation) Act 1978
GMC:	General Medical Council
GP:	General Practitioner
HSC:	Health and Safety Commission
HSE:	Health and Safety Executive

HSWA:	Health and Safety at Work Act 1974
IIAC:	Industrial Injuries Advisory Council
MAT:	Medical Appeal Tribunal
MEL:	Maximum Exposure Limit
MFOM:	Member of the Faculty of Occupational Medicine of the Royal College of Physicians
MSC:	Manpower Services Commission (abolished 1988)
NHS:	National Health Service
NI:	National Insurance
NIOSH:	National Institute for Occupational Safety and Health (USA)
OES:	Occupational Exposure Standard
OH:	Occupational Health
OHAC:	Occupational Health Advisory Committee
POM:	Prescription Only Medicine
RCN:	Royal College of Nursing
REA:	Reduced Earnings Allowance
RIDDOR:	Reporting of Injuries, Diseases and Dangerous Occurrences Regulations 1985
SMP:	Statutory Maternity Pay
SSAT:	Social Security Appeal Tribunal
SSP:	Statutory Sick Pay
TURERA:	Trade Union and Employment Rights Act 1993
UKCC:	United Kingdom Central Council (for Nurses, Midwives and Health Visitors)

Table of Cases

Table of Statutes

Table of Statutory Instruments

General Introduction

Law and ethics

Doctors and nurses are subject to the law and to the courts. As professionals, they are also ruled by ethical principles which may impose more onerous duties. In the United Kingdom the regulation of the professions is delegated by Parliament to professional bodies, like the General Medical Council (GMC) for doctors (Medical Act 1983), and the United Kingdom Central Council (UKCC) for nurses, midwives and health visitors (Nurses, Midwives and Health Visitors Act 1979). The decision whether an individual has behaved so disgracefully that he or she is unfit to continue as a doctor or nurse is granted to committees of these bodies. Appeals to the courts are only likely to succeed if the professional conduct committee has failed to hold a fair hearing, or has reached a decision which the judges consider so glaringly unreasonable that it must be beyond the committee's remit. Mr Singh, a senior nurse, was struck off the register of nurses in 1984 because he had given intravenous injections without a Certificate of Competence from his employing health authority (*Singh* v. *UKCC*). He had not contravened any legal provision, nor had any patient been injured. The decision of the Professional Conduct Committee of the UKCC reflected the concern of the nursing profession that nurses should not accept delegated tasks from a doctor without proper training (the Certificate of Competence was regarded at the time as necessary proof of such training). A High Court judge refused to interfere:

> 'Whether on the facts proved the conduct of the appellant amounted to misconduct is a matter for the members of his profession, and an appellate tribunal would be very slow indeed to substitute its own opinion for that of experienced members of the profession ...'

English law is found in the decisions of courts, in Acts of

1

Parliament, and in statutory regulations made by authority of Parliament. Ethical rules are more difficult to discover. Both the medical and nursing professions now publish guidance to their members in fairly general terms. The latest edition of the GMC's 'Bluebook' is dated 1993. The UKCC has produced a number of booklets, ranging from a general Code of Conduct (1992), to special guidance on confidentiality and the administration of medicines. The Faculty of Occupational Medicine of the Royal College of Physicians in 1993 published revised advice on ethics for doctors practising in occupational health. None of these codes is directly binding in law: they are an indication to the professions of the attitudes of other professionals who may be called upon to sit in judgment on their professional practice. This is not to underestimate their importance. The loss of the right to practise one's skills is a sanction greatly to be feared.

It is important to realize that neither the British Medical Association (BMA) nor the Royal College of Nursing (RCN) has any statutory function. They are 'trade unions' whose job is to represent their members. Nevertheless, both bodies are active in ethical debate, as the current dilemma over the treatment of AIDS patients illustrates. Disagreement is possible. The GMC has ruled that a doctor may break the confidence of a child too young to be able to appreciate his medical condition and therefore to give consent to treatment. The BMA, on the other hand, has argued that even a child's secrets must be respected, lest he be deterred from seeking help. In 1993 the BMA published a comprehensive survey of medical ethics: *Medical Ethics Today*. The RCN in its turn has produced a Code of Professional Practice for occupational health nurses (1987).

Law and ethics may conflict. The law says that confidence dies with the individual, but many professionals consider it unethical to reveal clinical details after the patient's death. The law allows the testing of anonymous samples of blood, such that the donor of the sample cannot be told of any defect. Is this ethically sound? The Helsinki Declaration on experiments on human subjects states that experimental treatment of a therapeutic nature, if approved by an ethics committee, does not always require the subject's consent. English law permits no such exception to the need for informed consent. Lawyers are not competent to advise on ethics, but the courts will not sanction unlawful behaviour simply because it is regarded as ethical, just as the professions will not approve unethical behaviour simply because it is lawful. Fortunately, in most cases law and ethics agree.

There are now international ethical principles established in the field of occupational medicine. The International Commission on Occupational Health in 1992 published an International Code of Ethics for Occupational Health Professionals. This gives three basic principles:

(1)   Occupational health practice must be performed according to the highest professional standards and ethical principles. Occupational health professionals must serve the health and social well-being of the workers, individually and collectively. They also contribute to environmental and community health.
(2)   The obligations of occupational health professionals include protecting the life and the health of the worker, respecting human dignity and promoting the highest ethical principles in occupational health policies and programmes. Integrity in professional conduct, impartiality and the protection of the confidentiality of health data and of the privacy of workers are part of these obligations.
(3)   Occupational health professionals are experts who must enjoy full professional independence in the execution of their functions. They must acquire and maintain the competence necessary for their duties and require conditions which allow them to carry out their tasks according to good practice and professional ethics.

This is bland and uncontroversial. General comments which may be made are to note the emphasis on prevention and health promotion, the expansion of the role of occupational health professionals into the field of environmental protection of the wider community and the need for 'a programme of professional audit of their own activities in order to ensure that appropriate standards have been set, that they are being met and that deficiencies, if any, are detected and corrected'.

The sources of English law

In modern times, law is found in statute and in precedent. Because English law has never been brought together into one code, the judges are still competent to make legal rules without reference to Parliament. This law, made by the judiciary, case by case, brick on brick, is known as the common law. Much of the law of contract

and tort (civil liability for unlawful acts) is still in this form.
Criminal law, on the other hand, is nearly all enacted in statute.
Judges, therefore, fulfil two functions in our system. They declare
and develop the common law. They interpret the meaning of Acts
of Parliament.

As in the medical profession, the opinion of those at the top of the
hierarchy is more respected than that of juniors. So, decisions of the
Judicial Committee of the House of Lords, the highest court, are
binding on all other courts. Decisions of the Court of Appeal are
binding on all courts other than the House of Lords. But even the
most senior judges must yield to the will of elected representatives
of the people. In our unwritten constitution, Parliament can overturn every judgment by a simple statute.

Parliament has no time to consider every detail of complex
legislation. Its practice is to establish broad principles in a parent
Act, giving power to a delegate, often a Minister, to make regulations which will be laid before it, and in a few cases will need its
affirmative approval. The Health and Safety at Work Act 1974
provides that an employer must do that which is reasonably
practicable to ensure that his employees are reasonably safe;
delegated legislation in the form of statutory instruments made
under the authority of the Act lays down detailed provision for
safety representatives, first aid, substances hazardous to health and
so on.

Legislation, both primary and delegated, carries binding legal
sanctions. It tends to be interpreted literally by English courts. A
statute which talks about 'employees' will not apply to self-
employed contractors. Laws which give compensation for
'accidental injury' will not cover cases where an employee has
contracted a disease over a long period of exposure. If legislation is
subject to decades of this literal approach, it becomes over-complex
and impenetrably obscure.

Draftsmen strive to express themselves more and more clearly,
but tend to lose the policy in a plethora of technical vocabulary.
Within the last twenty years, especially in the field of employment
law, it has become popular to attach Codes of Practice to Acts of
Parliament, to which the Act directs courts and tribunals to refer as
an aid to interpretation. Though these Codes are not in themselves
the law, to fail to follow their advice will be *prima facie* evidence of a
breach of the law. Examples are Approved Codes of Practice under
the Health and Safety at Work Act and the ACAS Codes under the
Employment Protection (Consolidation) Act 1978. Amendment of a
Code is simpler than altering legislation and judges can be more
flexible in their interpretation.

Criminal and civil law

The criminal law is concerned with the punishment of those who offend against society as a whole. Criminal prosecutions are brought by public officials such as the Crown Prosecution Service (Procurator Fiscal in Scotland) and the Health and Safety Executive. The defendant has to bear the costs of his defence, unless he qualifies for legal aid. If a defendant is convicted, he will be sentenced to some form of penalty, like a fine, imprisonment or community service. The money paid in fines goes to the courts, not to the victims of crime. Compensation is a function of the civil law. The plaintiff (pursuer in Scotland), the individual harmed by an unlawful act, sues in the civil courts or tribunals for damages to make up for what he has suffered. He may also ask for a court order, like an injunction (interdict in Scotland), directing the defendant to return to legality, breach of which will be a contempt of court. The plaintiff has to finance his own civil action, unless poor enough to qualify for legal aid. Up to now, the English and Scots legal professions have set their face against contingent fees which are paid to the lawyer out of any damages awarded to his client. However, it will soon be possible for a solicitor in England to represent his client on terms that he will waive his fee if he is unsuccessful (the Scots already permit this) and English solicitors will be able to charge a higher fee if successful in litigation. The winner of an action in the civil courts will almost always obtain an order that the loser must pay his costs.

The separation of punishment and compensation is not absolute. Criminal courts are empowered to order the convicted criminal to pay small amounts of compensation. Victims of violent crime can claim compensation, financed by taxation, from the Criminal Injuries Compensation Board. Those who flout an injunction granted by a civil court may be jailed for contempt. Also, many incidents give rise to both civil and criminal proceedings. For example, a company fails to provide protection for employees working with asbestos. It is prosecuted, convicted and fined in the criminal court for breach of regulations made under the Health and Safety at Work Act, a criminal statute. One of the employees concerned is diagnosed as suffering from lung cancer and asbestosis. He claims a disablement pension from the DSS and sues his employer in a civil court for damages for the torts of negligence and breach of statutory duty. The criminal penalty will be paid by the company, but the compensation will be paid by insurance, state-administered in the case of the pension and privately organized in the case of the award of damages.

The geographical extent of the law

The United Kingdom consists of England, Wales, Scotland and Northern Ireland. England and Wales have the same law, but Scotland has a separate legal system and different procedures. As a general rule, Acts of Parliament relating to health and safety, employment law and equal opportunities apply throughout Great Britain (which includes Scotland but excludes Northern Ireland). Northern Ireland is at present governed directly from Westminster. Legislation is usually extended to the Province by statutory instrument.

The Health and Safety at Work Act (Application outside Great Britain) Order 1989 extends the application of the Health and Safety at Work Act to offshore oil and gas installations, pipeline work, offshore construction, diving operations etc. Safety statutes do not otherwise protect those who work abroad: they will have to rely on local regulations. The employment protection legislation also applies to offshore workers, but not to those who ordinarily work in foreign countries. In *Todd* v. *British Midland Airways* (1978), an airline pilot who flew from Britain but spent 53 per cent of his time abroad was held ordinarily to work here. A national of any Member State of the European Community must be allowed to work in any other Member State and will be entitled to the same social security benefits as the 'locals'. Medical and nursing qualifications obtained in any Member State must be recognised in all other Member States.

Where a worker injured abroad seeks compensation from his employer, who is unwilling to defend the case here, the UK court might decide to refuse to entertain the case because the foreign court is more appropriate. In one case of an industrial accident which occurred in Scotland, the English court decided that the injured workman must sue in the Scottish court, because all the witnesses were in Scotland (*MacShannon* v. *Rockware Glass* (1978)).

The law of the European Union

Since the meeting of Heads of State of EC countries in Maastricht in 1991 there are four European Community treaties: the European Coal and Steel treaty, Euratom, the European Community treaty and the Treaty on European Union. The treaty which founded the European Economic Community, the Treaty of Rome, sought to create a common market between the Member States by providing for the free movement of goods, persons, services and capital and restraining anti-competitive measures like monopolies and restrictive practices. It was incorporated into United Kingdom law by the European Communities Act 1972. The following twelve countries are now members of the Union: Belgium, France, Italy, Luxembourg, Netherlands, Germany, Denmark, Eire, United Kingdom, Greece, Spain and Portugal. The Rome treaty was substantially amended by the Single European Act (1986) and more recently by the Treaty on European Union agreed at Maastricht in 1992. Because the European Union has no separate legal existence, the term 'European Community' is still in use.

The power to make new laws is given to the Council of Ministers and to the European Commission. Each Member State sends one Minister to the Council meetings. Sometimes the meeting is of Prime Ministers, sometimes of Agriculture Ministers, and so on. The bureaucracy of the Community is the European Commission which is situated in Brussels. The judicial power is conferred on the Court of Justice of the European Community (ECJ) in Luxembourg. Each Member State has one judge on the court. Its principal task is to interpret the treaties and secondary legislation made with their authority. Where the European Court makes a ruling on Community law, the courts of Member States must recognise and enforce it, and there is no appeal from its decisions. The directly elected European Parliament has up to the present had only a consultative and debating function, and its powers still fall far short of a legislative role, though the Maastricht treaty gives it a power of veto in limited areas.

Secondary legislation takes the form of regulations, directives and decisions. Regulations are mandatory. They have the force of law throughout the Community without the need to be ratified by the legislatures of the Member States. Directives are 'binding as to the result to be achieved', but leave the choice of method to the states concerned. They therefore require domestic implementing legislation. The Consumer Protection Act 1987 was passed to give effect to the principles laid down in the Product Liability Directive. It is not unknown for states to drag their feet. More than once has the United Kingdom been taken to the European Court by the Commission for failure to implement a directive. In *Marshall* v. *Southampton HA* (1986), the ECJ held that an individual employed by a Government could sue that Government under the provisions of an unimplemented directive as though it had been implemented, a privilege not available to non-Government employees. Employment in the National Health Service was held to be Government service. The European Court further developed the law in *Francovitch* v. *Italy* (1990) when it held that an individual may in some circumstances sue a Government for damages for failure to enact a directive within the specified period.

Decisions are rulings given by the Commission in individual cases and may be addressed to a state, an organisation or an individual. They are binding only on the individual addressed. Recommendations are persuasive, but not legally binding. Salvatore Grimaldi was born in Italy, but had worked for a long period in mining and construction in Belgium. He was diagnosed as suffering from an osteo-articular or angioneurotic impairment of the hand (Dupuytren's contraction), which he claimed was an occupational disease caused by the use of a pneumatic drill. This was not a prescribed disease under Belgian law, and Grimaldi was refused social security compensation. He appealed to the European Court.

The European Commission had made recommendations in 1962 and 1966 setting out a 'European schedule of occupational disease', including 'illness for over-exertion of the peritendonous tissue', and calling on Member States to introduce legislation granting compensation to those workers affected by such diseases and also to those able to prove that their disease was caused by work but unable to take advantage of domestic law because the disease was not prescribed. The European Court held that recommendations could not confer rights directly on individuals, but should be taken into consideration by national courts when interpreting domestic legislation, e.g. in cases of ambiguity (*Grimaldi* v. *Fonds des Maladies Professionelles* (1990)).

The European Treaties and laws made thereunder deal primarily with matters relating to the establishment of free trade. Much of our law is unaffected by a European dimension. The EC Treaty is not concerned with the law of theft, the grounds for divorce, the validity of wills or the need to obtain the consent of a patient to medical treatment. But a free market demands that no enterprise should be able to obtain an unfair advantage by ignoring essential safety measures imposed by law on its competitors. Every producer must start from the same base-line. Member States are unwilling to pass laws to protect the health of workers or the community which demand costly expenditure, if other States are permitted to maintain nineteenth century practices.

The EC Treaty, Art. 117, anticipated the need to deal with social as well as economic problems: 'Member States agree upon the need to promote improved working conditions and an improved standard of living for workers ...'. Article 118 provides that the Commission has the task of promoting close cooperation between Member States in the social field, particularly in matters relating to employment, labour law and working conditions, basic and advanced vocational training, social security, prevention of occupational accidents and diseases, occupational hygiene, and the right of association and collective bargaining between employers and workers. In 1974, an Advisory Committee on Safety, Hygiene and Health Protection was established to assist the Commission. This was instrumental in helping to draw up Action Programmes on Safety and Health at Work in 1978 and 1984. Some of the main aims of the programmes were as follows:

(1) To establish a common statistical methodology in relation to accidents at work and work-related diseases.
(2) To promote research.
(3) To standardise terminology for toxic substances, and harmonise exposure limits.
(4) To develop preventive and protective action against carcinogens.
(5) To establish principles relating to the protection of workers from exposure to harmful chemical, physical and biological agents at work.
(6) To establish common warning signs for the information of workers.
(7) To establish limit levels relating to noise and recommend steps to protect workers against noise.
(8) To establish a common methodology for the monitoring of pollutants in the work-place and to promote methods for the assessment of individual exposure, especially of women and adolescents.
(9) To undertake a joint study of the principles and methods of industrial medicine.
(10) To establish principles relating to major accidents.

The adoption of directives to implement these initiatives was hampered by the necessity to obtain unanimity among Member States. It was possible for only one State to veto any measure in the Council of Ministers. In 1987, however, the Single European Act came into operation. The countries of Europe agreed to establish a truly common market, with the abolition of all barriers to trade, by 1992. The resultant amendments to the Treaty of Rome include Art. 118A. Under this Article, it is now possible to adopt directives laying down minimum health and safety standards which exceed recognised standards in Member States. Qualified majority voting gives larger nations more votes than smaller ones, but allows even one of the 'Big Four' (France, Germany, Italy and the United Kingdom) to have legislation forced on it by the other Member States.

Similar principles apply to the enactment of directives in the field of consumer safety (Art. 100A). The EC Council may by a qualified majority approve standards for electrical and other goods. In this area, the standards are both minimum and maximum, lest states attempt to protect the home producer by keeping out goods which do not conform to unnecessarily high standards. The long-term solution is to establish European standards to replace those of the individual countries, and European organisations have been created to bring this about, Comité Européen de la Normalisation (CEN) and Comité Européen de la Normalisation Electronique (CENELEC) (for electrical apparatus).

The Single European Act (Article 130) for the first time included specific provisions allowing the Council of Ministers to legislate by a qualified majority in the area of environmental protection by, for example, setting minimum standards for toxic emissions into the atmosphere and water purity. States are permitted to set higher standards for themselves, unless this conflicts with the free market. In *EC Commission* v. *Denmark* (1989) a Danish law that beer and soft drinks could only be marketed in reusable containers for which a deposit must be charged was upheld by the European Court, even though it to some extent discriminated against foreign producers, because it reduced the quantity of litter damaging the environment.

The pace of change has been considerably expedited by the new system. Between 1970 and 1985 only six health and safety at work Directives were adopted by the European Council. In July 1987, however, a third Action Programme on safety, health and hygiene at work was adopted. A long list of measures was proposed by the Commission, including fifteen new Directives. By 1989 a 'Framework' Directive for the Introduction of Measures to Encourage Improvements in Safety of Health of Workers was approved by the Council of Ministers. This has been enacted into UK law by the Management of Health and Safety at Work Regulations 1992 which require employers to assess risks to employees, provide them with health surveillance, give them information and training and appoint competent persons to supervise a safe system of work.

Shortly after, five 'Daughter Directives' containing more specific provisions about health, safety and welfare provision in the workplace (heating, lighting, ventilation, cleanliness and so on), machinery and work equipment safety, personal protective equipment, visual display units and the handling of heavy loads were passed. These are now incorporated into UK law as the Workplace (Health, Safety and Welfare) Regulations 1992, Provision and Use of Work Equipment Regulations 1992, Personal Protective Equipment at Work Regulations 1992, Health and Safety (Display Screen Equipment) Regulations 1992 and Manual Handling Regulations 1992. All these regulations are discussed in detail in Chapter 5.

1992 was designated as the European Year of Safety, Hygiene and Health Protection at Work by the Council of Ministers. The four main themes for the year were clean air at the workplace, safety at work, well-being at work and measures against noise and vibration. The Community's particular priorities were small firms, young people in training and the high risk sectors of construction, agriculture, mining and fishing.

Although the safety of workers has been an important part of European Community policy since the foundation of the Coal and Steel Community in 1951, and the Treaty of Rome provides specifically for equal pay for men and women at work (see Chapter 9), the use of Community law to lay down minimum rights for workers in other areas is far more controversial. Article 100A permits legislation on working conditions if it is necessary to ensure the effective functioning of the common market, and measures to protect workers affected by collective redundancies, transfer of the undertaking in which they are employed and the insolvency of their employer have been enacted under this power. However, provisions relating to the free movement of persons or the rights and interests of employed persons must be agreed by all Member States: every State has the power of veto.

Conflict has arisen between the majority of European governments who regard the protection of the worker as an important aim for social legislation and the free marketeers in the UK government who prefer to allow market forces to determine workers' rights unfettered by what they regard as artificial barriers to economic growth.

In December 1989 eleven Member States signed a Community Charter of Fundamental Social Rights of Workers (the Social Charter). This is largely a declaration of policy, without direct legal effect. The United Kingdom government stood alone in refusing to agree to the principle that workers' rights should be enshrined in European law, and did not sign the Charter. The Charter's provisions are expressed in general aspirational terms. It deals with minimum wages, working hours and holidays, rights of association and the right to negotiate and conclude collective agreements.

At the meeting of governments at Maastricht in December 1991, the majority of Member States agreed that they wished to adopt further measures to pursue the aims of the Social Charter. Again, the UK government was in a minority of one. The compromise eventually reached (the Maastricht Protocol) was that all twelve Member States agreed that eleven States (all except the UK) might have recourse to the institutions, procedures and mechanisms of the Treaty of Rome for the purposes of taking the acts and decisions necessary to give effect to the new agreement. The UK government will not participate in this process, though it will not interfere with the eleven. There will therefore be two kinds of social policy legislation in the Community:

(1) 'Pre-Maastricht' legislation, in which the UK government will play a full part, where the rights proclaimed in the Charter already fall within the legislative competence of the Community, e.g. improvements in the working environment as regards health and safety of workers (Article 118A) and equal pay for men and women (Article 119);
(2) 'Maastricht Protocol' legislation, to which the UK government will not be party, and which will not be binding on the United Kingdom.

The political and social problems created by this two-tier system are many. In particular, its implementation may result in the British worker becoming the poor man of Europe, with significantly lower wage rates and reduced legal protection against exploitation by the employer, although increased trade through the production of more competitive goods may increase general prosperity. There are already signs that multi-national companies look favourably at siting their plants in the UK to take advantage of lower rates of pay.

The legal problems are formidable. It is not easy to differentiate between the various types of legislation. Take, for example, the draft directive on working hours, which proposes minimum daily and weekly rest periods and annual paid holidays of a minimum length. If, as the Commission has argued, this falls within Art. 118A as a health and safety measure it can be passed by a qualified majority without the consent of the UK government and must then be enacted into UK law. If it is to be classified as a 'workers' rights' measure, it may either fall within Article 100A, in which case it may only become law if all the Member States agree, or alternatively (but less likely) be regarded as an attempt to implement the Social Charter falling within the Maastricht Protocol and therefore not involving the UK at all. This issue is likely to be litigated in the European Court, thus delaying the implementation of the directive.

The Social Charter and the Maastricht Protocol emphasise the principle of subsidiarity, namely that the Community must recognise the differing social structures and the diversity of national practices of the Member States. There can be no Community-wide minimum fair wage, because of the many differences in cost of living, incidence of taxation and social security etc., but there can be the espousal of the concept of a legally guaranteed fair wage for each individual country.

The UK government has expressed concern that, at a time when it is pursuing a policy of deregulation on the home front with the aim of freeing industry from petty legal restrictions which deter growth, new and onerous duties are being introduced through membership of the European Community. 'Health and safety legislation could be brought into disrepute by introducing bureaucratic requirements which would not be complied with' (*Review of implementation and enforcement of EC law in the UK* (1993) Department of Trade and Industry). The DTI report recommends the designation of one of the Health and Safety Commissioners as especially concerned, as far as possible, to look after the interests of the small employer. The implementation of EC law is for the Member States, who must adjust the European principles to their own domestic institutions, but the essence of the principles must reach the statute book. In particular, the granting of exemptions to the self-employed and small businesses, unless permitted by the Directive, would be a breach of European law for which the government could be condemned by the European Court, though Art. 118A directs the Member States to be conscious of the needs of small and medium-sized undertakings.

International Labour Organisation

This was founded in 1919 by those nations who had been the victors in the First World War to bring together representatives of employers' and workers' organisations and also of governments of participating States. It holds international conferences annually in Geneva and acts as a focus for those who strive to raise standards of protection for workers, not merely in the field of health and safety, but in industrial relations in general. Most countries belong to the

International Labour Organisation (ILO) which is the oldest and most experienced international body concerned with the establishment of international labour standards.

The ILO adopts Conventions and Recommendations. Ratification of a Convention amounts to an undertaking that its provisions will be given legally binding force by a legislative enactment. Even then, a government may denounce a Convention at a later date, as the British Government has done with laws preventing women workers from being employed at unsocial hours (see Chapter 9). A Recommendation does not have to be ratified, but adoption by a government signifies that it will in the future be guided by the Recommendation if and when it decides to act. The ILO has produced a Convention and Recommendation relating to the provision of occupational health services (see Chapter 1).

Chapter 1

The Provision of Occupational Health Services

1.1 The development of occupational health services

The origins of occupational health provision lie in the heyday of the Industrial Revolution. Workers in the mills and factories, in common with all except the well-to-do, had no access to medical services because they could not afford them. Some benevolent employers, moved by the suffering of the masses, provided housing and medical services out of their profits; most did not. The nature of this provision was not in any sense connected with work-related disease or injury; it was general medicine for workers and their families such as is today provided by the general practitioner in the National Health Service. Workers still perceive the provision of medical and nursing services at work as a mark of a good and caring employer; it goes together in their minds with decent canteen facilities and a good working environment. On the other hand, now that the NHS gives everyone access to free medical treatment, it may be considered wasteful for there to be duplication of treatment facilities, other than to provide first aid in an emergency. This argument might be more easily sustained if the NHS were not under constant financial pressure. Also, if the provision of physiotherapy at the workplace saves the worker having to take a day off a week to attend the hospital it may be of financial benefit to the employer and reduce the burden on public funds. Increasingly, employers in the private sector see the health of their key workers as a business asset to be maintained with medical and nursing assistance in the same way as engineers maintain machinery, though in practice regular health surveillance of such workers is often contracted out to private health organisations like BUPA.

Other developments which contributed to the growth of occupational health (OH) services were various Acts of Parliament passed to give the employee a right to compensation against his employer (beginning with the Workmen's Compensation Act 1897), long since transferred to the Welfare State under the industrial injuries legislation, and to protect the consumer against risks caused by the ill-health of workers in, for example, the food processing and transport industries. The principal motives behind the introduction of medical monitoring by occupational health professionals in response to these measures were to protect the employer against legal action and the public against injury, rather than to care for the welfare of the workers, though the genuine concern for their employees of pioneer companies like Chloride and Pilkington's must also be acknowledged. Other factors were the increase in statutory regulations to protect the munitions workers during the First World War and the need after both World Wars to help the disabled to find and maintain suitable employment.

After the Second World War there were several official reports on provision for occupational health, including the Dale Report in 1951 and the Porritt Report in 1962. The Robens Committee on Health and Safety at Work, reporting in 1972, stated that in their understanding occupational health included two main elements – occupational medicine, which is a specialised branch of preventive medicine, and occupational hygiene, which is the province of the chemist and the engineer engaged in the measurement and physical control of environmental hazards. 'Clearly these two elements must be closely integrated, since the basis for environmental control must be derived from the medical assessment of risk.' The Committee placed the greatest stress on their fear that the employment of large numbers of doctors and nurses in the workplace would be a wasteful duplication of the general practitioner service. They were largely in agreement with the view of the Government that : 'In the field of occupational health the working environment is of predominant importance, and it is engineers, chemists and others rather than doctors who have the expertise to change it'.

The Health and Safety Commission (HSC) in 1977 produced a wide-ranging discussion document: *Occupational Health Services – The Way Ahead*. This highlighted the problem of providing services for workers in small organisations. It explored various ways of promoting cooperation between employers, like the establishment of group industrial health services (Slough is a well-known example) to which small companies could subscribe according to the number of their employees, or the 'leasing' of spare capacity in a large organisation to other employers in the locality.

The Health and Safety Executive (HSE) in 1982 published a booklet entitled *Guidelines for Occupational Health Services*, which gave practical guidance on the functions, staffing and operation of OH services. This stressed that each organisation has its own needs. The number of employees, the number of locations, the number and severity of potential hazards, any statutory requirements for health surveillance, and the availability of and distance from NHS facilities must all be taken into account.

In 1983 the Select Committee on Science and Technology of the House of Lords chaired by Lord Gregson reviewed the future provision of occupational health and hygiene services. It defined occupational health as the physical and mental well-being of the workers and occupational hygiene as the control of physical, chemical and biological factors in the workplace which may affect the health of the worker. The Gregson Committee perceived the main aim of an occupational health service as the promotion of the health and safety of those employed at the workplace. Occupational medicine was described as 'a branch of preventive medicine with some therapeutic functions'. No full survey of occupational health services had ever been undertaken, but what research had been done revealed that at that time (1976), full-time medical and nursing personnel were concentrated in large industries, as might be expected. Many large companies relied on part-time medical advisors who might be local general practitioners (GPs). Few of these had special training in occupational medicine: 87.6 per cent of firms, employing 36 per cent of the workforce, had no medical service apart from first-aiders (*Occupational Health Services – The Way Ahead* (1977), *op. cit.*).

The Committee concluded that more provision was needed in small firms. They put considerable emphasis on preventive medicine:

> 'Early detection of hazards of work and the timely adoption of preventive measures will not only alleviate individual suffering: they will lighten the financial burden which sickness imposes upon the State. There are also sound business reasons for ensuring that a workforce remains healthy. A healthy worker is a more efficient worker: absenteeism is lower and productivity higher.'

The costs of the service should continue to be borne by the employers in reflection of their general duty under the Health and Safety at Work Act. However, Gregson was not in favour of imposing a legal obligation to provide an occupational health service. The Committee thought that a non-statutory Code of Practice should be drawn up and monitored by the Employment

Medical Advisory Service (EMAS). Tax incentives could be conferred on those who implemented the Code, and insurance companies might take it into account in fixing premiums. General practitioners should be encouraged to extend the occupational health side of their activities and to acquire additional qualifications. Occupational health nurses should be the first point of contact between the patient and other sources of referral. Trade unions and employees should be given more opportunity to have a voice in the management of occupational health services.

So far, there has been no significant move towards either a voluntary or a statutory Code of Practice. Meanwhile there have been international developments. As long ago as 1962, the European Commission recommended that a statutory obligation to provide an occupational health service should be introduced at least for large employers (as has been shown, this would not represent much of a change in this country where most large concerns already have such a service). In June 1985, the International Labour Organisation (ILO) adopted a Convention (No. 161) and a supporting Recommendation (No. 171) on Occupational Health Services. The Convention defines occupational health services as 'services entrusted with essentially preventive functions and responsible for advising the employer, the workers and their representatives in the undertaking on:

(1) the requirements for establishing and maintaining a safe and healthy working environment which will facilitate optimal physical and mental health in relation to work; and
(2) the adaptation of work to the capabilities of workers in the light of their state of physical and mental health.'

It should be noted that the Convention covers occupational hygiene and ergonomic services as well as medical and nursing services. Signatories to it will have to formulate, implement and periodically review a coherent national policy on occupational health services and to develop progressively occupational health services for all workers such as are adequate and appropriate to the specific risks of the undertakings. The UK Government has sought the advice of the HSC on whether the UK should ratify and/or accept Recommendation 171, which is in similar terms to the Convention, but would not, if implemented, carry the same mandatory legal force as the Convention, were the Convention to be adopted here. The HSC has advised that no decision should be taken at this stage.

If this country were to ratify the Convention, legislation would eventually be needed. An additional duty would have to be imposed on employers by amendments to the Health and Safety at Work Act whereby they would be compelled to provide an adequate and appropriate occupational health service, as defined in the legislation, or be guilty of a criminal offence. Further legal provisions would be needed to implement specific requirements. The legislation would not have to come into immediate effect, but would commit us to a process of progressive development. The Recommendation could be accepted only in part; there would be a moral though not a legal obligation to implement any provisions which had been accepted. However, acceptance of the Recommendation would not require any major changes in our law.

As the Convention is under discussion, it may be worthwhile to examine its structure, especially as it demonstrates the trend of the international community's thinking on occupational health services. Important aspects are as follows:

(1) It employs legal sanctions, rather than the voluntary approach which has so far prevailed in the UK.
(2) It adopts a multidisciplinary approach, regarding the doctor and the nurse as part of a team which also includes the hygienist and the ergonomist.
(3) It contemplates that there shall be recognised qualifications for personnel providing occupational health services.
(4) It requires the involvement of the workers themselves in the management of the service.
(5) It sees the function of the service as essentially preventive; treatment is confined to first aid and emergency treatment.

As has been seen, the movement in occupational health has been away from treatment but towards prevention. If an employer wishes to provide 'private treatment' services in addition to the NHS because he thinks them economically worthwhile he may do so, but this is less important than the identification of work-related hazards and the steps taken to protect the workers against them. A Joint ILO/World Health Organisation Committee in 1950 wrote this:

'Occupational health should aim at the promotion and maintenance of the highest degree of physical, mental and social well-being of workers in all occupations; the prevention among workers of departures from health caused by their working conditions; the protection of workers in their employment from risks resulting from factors adverse to health; the placing and maintenance of the worker in an occupational environment adapted to his physiological and psychological equipment.'

The UK Government's response to the Gregson report was published in 1984 (Select Committee on Science and Technology). It enthusiastically welcomed the conclusion that 'the responsibility for occupational health and hygiene services should lie largely with individual employers'. It doubted whether GPs could provide services as an integral part of primary care without reducing efficiency. The Government made clear its determination not to provide OH services out of public funds.

At the same time, the HSC agreed that there should be a review of the participation of GPs in occupational health practice. The Commission supported the provision of a training and qualification scheme appropriate for doctors working for only a small part of their time in industry. It recognised 'the important role of trained occupational health nurses ... which is frequently misunderstood and ... could be widened in scope'.

In 1986 the HSC issued a statement of policy in response to the Gregson Report. It declined at that stage to formulate a Code of Practice. Instead it has initiated a programme of action. This will include:

(1) the preparation of guidelines for employers on such matters as the benefits and availability of OH services;
(2) a publicity campaign especially aimed at small firms about the appropriate use of OH services;
(3) the encouragement of new projects by the Industry Advisory Committees (organisations set up within particular industries, under the aegis of the HSE, to make recommendations to the HSC);
(4) the promotion by the HSE of conferences and seminars for the exchange of practical information about the provision of OH services;
(5) liaison with training bodies to promote the training of OH specialists and make managers more aware of health and safety;
(6) the improvement of co-ordination with the National Health Service;
(7) co-ordination between the organisations involved in the provision of OH services, including larger employers, public and professional bodies, academic departments, group services and independent consultancies.

Many health professionals working in occupational health have criticised what they see as a lack of resolve in the HSC. Until the employment of specially trained OH personnel is at least strongly recommended by official bodies, they see employers, especially small firms, continuing to be unwilling to spend money on an efficient OH service. The employment of general practitioners on a part-time basis is popular with employers, but special training is necessary for these to appreciate the particular problems of occupational medicine.

A research report commissioned by the HSE entitled *Occupational Health Provision at Work* was published in 1993. The three-tier survey was based on 820 private sector and 100 public sector organisations and 912 employees. It was discovered that 65 per cent of private sector establishments have some occupational health measures (defined as any action which could prevent work-related ill-health), as compared to almost 100 per cent in the public sector. 89 per cent of the workforce is employed in establishments which have some occupational health measures. 34 per cent of all employees work in organisations which employ a doctor, either full-time or part-time, and 53 per cent have access to some health professional. The employee/doctor ratio is 212:1. In just under half of the establishments using health professionals in the private sector there was at least one person with specialist OH qualifications. This rose to 74 per cent in the public sector. The employees when asked to report health problems caused by work most frequently mentioned back and other physical strains and stress. Also mentioned were headaches and eye strain, deafness, skin problems and asthma.

Since 1990 there have been two major developments which strengthen the position of the specialists in occupational health. The first was the publication of the White Paper, *The Health of the Nation*, in 1992. Successive governments since the inauguration of the National Health Service after the Second World War had come to realise that spending on health care must be contained. The creation of an internal market by separating the authorities who provide health care from those who purchase it was one strategy to try to secure better value for money. Another is to try to encourage the population to take care of their own health. The White Paper demonstrates the Government's commitment to preventive medicine. It selects five key areas in which national targets are fixed. These are coronary heart disease and stroke, cancers, mental illness, HIV/AIDS and sexual health, and accidents.

The White Paper emphasises the importance of a healthy workplace and proposes the setting up of a task force to examine and develop activity on health promotion in the workplace. It also encourages the NHS to set an example to other employers to show what can be achieved. The NHS Management Executive has set up a task group of NHS managers, Health Education Authority representatives and professionals to review the way in which the NHS promotes the health of its own employees. This concentration on preventive medicine can only increase the scope for the trained occupational health professional.

The second development has been the increase in the numbers of regulations governing health and safety at work, particularly those originating in European Community Directives. Many of these are designed to prevent long term injury to health, as compared to the prevention of accidental injury. Health professionals with the necessary training and expertise will be especially valuable to employers who need advice on the implementation of the regulations and the provision of health surveillance to ensure that the employees are not suffering adverse effects from their work. Perhaps the most important of these regulations are the Management of Health and Safety at Work Regulations 1992, implementing the EC Framework Directive. These oblige all employers, with minor exceptions, to make a suitable and sufficient assessment of the risks to the health and safety of their employees, and to those not in their employment, arising out of the conduct of their undertakings. Every employer shall ensure that his employees are provided with such health surveillance as is appropriate. The Approved Code of Practice advises that, at least in some instances, this will necessitate the services of 'an Occupational Health Nurse' or medical surveillance by 'an appropriately qualified practitioner'. Taken with the emphasis in the regulations on the need to employ competent persons, it would seem that the employment of health professionals with specialist qualifications in occupational health is gaining official recognition. An Occupational Health and Safety Lead Body has been established to develop vocational qualifications for health and safety practitioners (including OH physicians and nurses).

The Health and Safety Commission expresses one of its priorities as the establishment of the key points of attack in improving occupational health and identifying the extent of occupational ill-health, and the taking of appropriate action to exploit the linkages between occupational health and the Government's 'Health of the Nation' initiative. 'The assessment and management of health risks – the central requirement of the various regulations – are often more complex or involve greater uncertainty than for occupational safety risks. Targeted guidance on assessment and management, and on selecting expert advice, will be needed by employers and employee representatives, as well as by health and safety inspectors, as an essential tool to ensure effective action.'

The Commission plans to continue to give high priority to epidemiological research. A successful consultant-based scheme for the surveillance of occupational lung disease is to be continued, and there is now a similar scheme for occupational skin disease. More attention is being paid to general practitioners, who may fail to identify the connection between a patient's work and his medical condition. The Health and Safety Executive in 1992 produced a booklet on occupational health for family doctors: *Your Patients and their Work*.

1.2   The legal obligations of the employer

The law imposes a number of specific obligations on the employer relating to the health of his workers and, more generally, in the Health and Safety at Work Act obliges him to ensure their health and safety so far as is reasonably practicable. So far, there is no specific duty in our law on the employer to provide qualified medical or nursing staff at the place of work. The *Health and Safety (First Aid) Regulations 1981* oblige employers to provide adequate and appropriate first aid equipment and facilities and an appropriate number of adequately qualified and trained persons to render first aid to his employees. The Approved Code of Practice recommends by way of a guide that at least one first-aider to every 50 employees is necessary, and more in areas of greater hazard. These numbers refer to employees present at work together at any time; where there is shift-working first-aiders may be needed for each shift. In high-risk areas like the manufacture of chemicals at least one first-aider trained not only in general principles but also in the particular risks of his employer's business should be available. Where a first-aider is not required, an appointed person must be available whenever employees are at work. This person must be capable of taking charge of a first-aid box and have instructions on how to call for medical assistance. A first-aid room with the necessary staff and equipment is only required when 400 or more employees are working at the same time or if the process is especially dangerous or in an isolated spot. The employer must tell his workers about first-aid facilities and where to find a first-aider. To be qualified as a first-aider under the statute, the employee must have been trained and received a certificate of qualification approved by the HSE.

Amendments and additions to the Approved Code of Practice came into force in 1990. More guidance is given on appropriate first aid materials and more extensive provision is made for training and for refresher courses (every three years).

It will be noted that there is no obligation in these regulations to have a qualified doctor or nurse regularly on the premises, though an employer who has an OH service will not have to comply with the detailed provisions of the Approved Code of Practice if he is in effect providing a service which reaches standards as least as high. However, when workers are employed in especially hazardous processes, like working with asbestos, lead or ionising radiations, statutory regulations may require regular health surveillance, which will demand the presence of medical personnel at least to supervise regular examinations. The introduction of the Control of Substances Hazardous to Health Regulations has substantially increased these requirements, because they not only extend the list of named substances (to include, for example, vinyl chloride monomer and benzene), but also oblige the employer to assess the hazards of all substances with which his employees have contact at work and to introduce regular surveillance where there is a reasonable likelihood that an identifiable disease or adverse health effect may occur from exposure to a hazardous substance (see Chapter 4).

1.3   Who pays?

In some industrial countries, the provision of an occupational health service is regarded as an important part of the Welfare State. In Italy, for example, the prevention of accidents and ill-health at work is one of the functions of the local health authorities. Though the NHS has in the past been seen primarily as a treatment service, preventive medicine has gained in importance, especially as soaring costs have placed intolerable strains on the Exchequer. The White Paper *Health of the Nation* (1992) demonstrates the new commitment to disease prevention and health promotion. Facilities for the treatment of non-emergency conditions at the workplace can with justification be charged to the employer. He may be willing to pay a doctor, dentist or nurse to treat his workers at the factory so that they do not have to waste working time travelling to the GP's surgery, though this will be resisted by the local GPs if they do more than treat emergencies and minor injuries.

But what of the preventive aspect of occupational medicine? The

State provides the Employment Medical Advisory Service, but the numbers of personnel are too small to be able to do more than lead and advise. Where statutory regulations oblige the employer to monitor the health of his workers he is compelled by law to pay the fees of an Appointed Doctor. The small employer will find it difficult to do much more than this, but there is evidence that employees in such enterprises are at greater risk than those in larger organisations. If the provision of an OH service for every employee is to be a practical possibility, either it must be taken over by the NHS or more cost-effective methods must be found. The Gregson Committee was not enthusiastic about an NHS takeover, principally because it found that the health service did not yet succeed in caring adequately for its own personnel, let alone any-one else's. In the four years since the publication of the first edition of this book, health authorities have improved their occupational health services and have in some cases successfully contracted out services to the private sector as a form of income generation. It would seem that the employer must continue to pay, but that costs could be reduced by a number of employers combining together, or a large enterprise offering spare capacity to small local firms. Current economic theory advises that market forces must be allowed to operate; if it is in the economic interests of employers to provide an OH service, they will eventually do so. Employers need to be shown how their business can benefit financially if they are to be persuaded to extend medical and nursing services at the workplace.

The Health and Safety Executive has sought to estimate the costs of accidents at work. There are some 1.6 million accidents resulting in injury every year, and some 750,000 workers suffering from ill-health caused or made worse by their work have had to take time off. This results in a loss of over 30 million working days and a cost to industry of almost £700 million. The overall cost of work accidents and work-related ill-health to employers is estimated to be between £4,000 million and £9,000 million a year. This esti-mation includes the £750 million cost of employers' liability insurance and the cost of recruiting and training replacements for injured workers, as well as property loss. This is equal to between 5 per cent and 10 per cent of gross trading profits. It does not include the costs of social security and the NHS, or the personal losses of the victims of accident and disease. Altogether, the total cost to society is estimated as between £10 billion and £15 billion a year (*The Costs of Accidents at Work* (1993)).

Advice from the Health and Safety Executive on occupational

health services (1989) advises that the benefits of such a service are:

(1) compliance with legal responsibilities;
(2) reduced labour turnover and increased efficiency;
(3) less sickness absence and fewer compensation claims through detection of health hazards and adoption of preventive measures;
(4) less waste of employees' work time through provision of on-site first aid and treatment facilities;
(5) improved general health through introduction of health promotion and education programmes;
(6) a better motivated workforce and higher calibre job applicants through showing that the employer cares about the health of its workers.

Most employed people work for employers whose workforce is too small for a comprehensive in-house occupational health service. But the specific needs of small and medium-sized firms can be met in a number of ways. A qualified occupational health nurse can be employed full-time, a part-time or visiting doctor may be appointed, the safety officers can have a key role in introducing and monitoring measures to control the working environment and in checking on their adequacy and effectiveness. Some firms become members of a group occupational health service which provides occupational health facilities on a shared basis to a number of local firms. Some large organisations with comprehensive in-house services are prepared to offer facilities to small local firms. This is especially true of NHS occupational health services based at local general hospitals. There are also independent organisations operating occupational health services on a consultancy basis, and services provided by private health insurers.

These services may be cost-effective. Several examples are given by the Health and Safety Executive. The first is an organisation requiring a high level of employee fitness which has a dispersed workforce of some 2500 people based at 26 sites, some of which are 30 miles or more from headquarters. Medical surveillance is carried out at headquarters by a part-time medical adviser, costing the employer both overtime payments and travelling expenses. The employment of a nurse to visit all the sites and refer to headquarters only those employees who have to see the medical adviser is more than compensated by reductions in overtime and travelling expenses for the workers. A second example is a chemical company employing over 1000 people and concerned about absence from work due to muscle and joint disorders. It arranges for a self-employed physiotherapist to attend three mornings a week. The employer benefits greatly from the reduction in absences from work from sickness and attending for treatment outside the workplace. Easy access to the physiotherapy service also assists early treatment which is often essential to a good outcome. Finally, a company using isocyanates requires health surveillance of the workforce. EMAS advises the company to contact a local GP. The GP notices other occupational health problems, such as skin conditions and stress-related symptoms. This leads to regular workplace visits, which improve the health status of the workforce and benefit manager/worker relationships.

1.4 The Employment Medical Advisory Service and Appointed Doctors

In 1833 and 1844 the Factory Acts required child workers to be examined by a local surgeon or physician to assess whether they were under the legal minimum working age (nine). The 1855 Factory Act conferred on these Certifying Factory Surgeons the task of certifying that young people were not incapacitated for work and of investigating industrial accidents. The doctor was independent both of the employer and of the employee, though paid by the factory owner. Later in the century, legislation required the Certifying Factory Surgeons also to investigate cases of industrial disease. In addition, employers were compelled to pay Appointed Surgeons to undertake regular medical examination of those working with specified substances like lead and phosphorus. The first full-time Medical Inspector of Factories was appointed in 1898.

Part-time medical practitioners have, therefore, conducted medical examinations in industry for nearly a century. They now operate under the supervision of full-time specialists in occupational medicine, the Employment Medical Advisory Service (EMAS) and are called Appointed Doctors (ADs). Their principal function is to undertake examinations of workers in the workplace and to assess their fitness for work when the employer is bound by statute to carry out regular health surveillance. About 3,000 examinations are carried out by EMAS and a further 90,000 by ADs every year. The ADs are appointed by the Health and Safety Executive through EMAS in respect of a particular company or companies and premises. EMAS is able to ask for evidence of occupational health qualifications and experience when making appointments, though these are not mandatory legal requirements.

The ADs are obliged to comply with both the clinical and administrative procedures set by EMAS; they are subject to inspection by Employment Medical Advisers. It is very important that proper clinical records are kept and statistical returns made to EMAS. If the work of the AD is unsatisfactory, his appointment may be revoked without a reason being stated or an explanation given. The employer has to pay fees for medical surveillance required by law.

The functions of EMAS are laid down in section 55 of the Health and Safety at Work Act. It undertakes the following responsibilities:

(1)   Advice to the inspectorate on the occupational health aspects of Regulations and Approved Codes of Practice (see Chapter 5).
(2)   Regular examinations of persons employed on known hazardous operations.
(3)   Other medical examinations, investigations and surveys. An Adviser has power to require an employer to permit him to carry out a medical examination of any employee whose health the Adviser believes may be in danger because of his work.
(4)   Advice to the HSE, employers, trade unions and others on the occupational health aspects of poisonous substances, immunological disorders, physical hazards, dust, and mental stress, including setting standards of exposure to harmful processes and substances.
(5)   Research into occupational health.
(6)   Advice on the provision of occupational health and first aid services.
(7)   Advice on the medical aspects of rehabilitation and training for and placement in employment.

EMAS is the field force of the Health and Safety Executive's Medical Division which is headed by the Director of Medical Services. EMAS employs both doctors (EMAs) and nurses (ENAs) as well as support staff. The Division also provides the secretariat for three of the Health and Safety Commission's Advisory Committees: the Occupational Health Advisory Committee (OHAC), covering wide-ranging questions in the field of occupational health, the Advisory Committee on Dangerous Pathogens (ACDP) and the Advisory Committee on Genetic Manipulation (ACGM). Scientists employed by the Division in the Health Hazards Assessment Branch assess hazardous agents and help to devise appropriate controls. There is also an Epidemiology and

Medical Statistics Unit (EMSU) and the Division commissions projects from outside bodies. The Data Appraisal Unit (DAU) has a general remit to assess and appraise information from a wide range of sources, published and unpublished.

1.5   The occupational health physician

The British Medical Association (BMA) defines an OH physician as 'a doctor who in relation to any particular workplace takes full medical responsibility for advising those working therein including contractors working on the site on all matters connected directly or indirectly with the work. These may have a bearing on health as it affects work and the effect of work on health including that of the public at large, either in general or as individuals'.

An OH physician must be a registered medical practitioner. He may also be an Associate Member of the Faculty of Occupational Medicine of the Royal College of Physicians (AFOM) or possess a Diploma in Industrial Health. Further experience and a dissertation enables a doctor to apply for membership of the Faculty (MFOM). There is as yet no legal requirement that the OH physician should hold any other qualification than ordinary registration. General practitioners often take up a post of OH doctor in an organisation on a part-time basis. Many part-timers are not Associates of the Faculty of Occupational Medicine, though it is possible to qualify for this by a period of experience in industry and an examination.

No formal arrangements exist for training in occupational health for medical students. Occupational medicine is the only medical specialty in Britain in which the cost of training is not defrayed from public funds. Obviously, the Faculty would prefer that a formal occupational health qualification be made a legal requirement, but at present there are insufficient training facilities available for the numbers required. The Faculty has now developed a short training course for physicians who do not need the full specialist qualification. In 1993 it decided to introduce an examination in occupational medicine which would standardise the assessment of all participants in these short courses. The syllabus will be expanded, and the examination will judge basic knowledge and competence in occupational medicine, with the award of a Diploma for successful candidates. This will remain quite separate from the AFOM/MFOM which will continue to be the route for those wishing to specialise in occupational medicine.

Another development has been the report of the Chief Medical

Officer's Working Group on Specialist Medical Training which is
likely to lead to changes in the training of occupational physicians,
with a curriculum which places more emphasis on competence
rather than purely factual knowledge.

The BMA sets out the duties of an occupational physician as
follows:

(1) *The effects of health on capacity to work*
(a) Advice to employees on all health matters relating to their
     working capacity. Examination of applicants for employment
     and advice as to their placement.
(b) Immediate treatment of medical and surgical emergencies
     occurring at the place of employment.
(c) Examinations and continued observation of persons returning
     to work after absence due to illness or accident, and advice on
     suitable work.
(d) Health supervision of all employees with special reference to
     those particularly vulnerable, like the disabled.

(2) *The effects of work on health*
(a) Responsibility for first aid services.
(b) Periodical examinations and medical supervision of persons
     exposed to special hazards.
(c) Advice to management regarding the working environment
     in relation to health, the occurrence and significance of
     hazards, the health aspects of safety, and statutory require-
     ments in relation to health.
(d) Medical supervision of health and hygiene of staff and facil-
     ities, with special reference to canteens, those working on the
     production of food and drugs for sale and those providing a
     service to the public.
(e) Health education for employees.
(f) Advice to health and safety committees.
(g) Liaison with other specialists working in the field.

1.6   *The occupational health nurse*

The first recorded occupational health nurse was Phillippa Flow-
erday who was appointed in 1878 by Colman's in Norwich. She
assisted the doctor at the factory and then visited sick employees
and their families in their own homes. Her work reflected the
treatment-based philosophy of the time and also the 'doctor's
helper' attitude to nursing staff. There has been no statutory

function for nurses comparable to that of the Appointed Doctors, so that OH nursing has been centred very much round the role of providing first aid in the workplace. The sympathetic nurse with time to listen and guaranteed confidentiality has also been a popular source of advice for such sociomedical problems as members of the worker's family drinking too much, overtiredness caused by stress, menopausal symptoms and so on. A survey undertaken on behalf of the Royal College of Nursing in 1982 showed that caring for the sick and injured and counselling were the two functions most often mentioned by OH nurses in describing their work.

Because nurses command for lower salaries than doctors, employers in times of prosperity were often willing to provide a nurse in the factory for the welfare of their workers. In more straitened times, they have been asking what the nurses can contribute that cannot be done equally well by first-aiders and welfare officers. Treatment is available free from the NHS; why should the employer duplicate it? The occupational health nursing profession is, therefore, now having to justify itself by demonstrating that nurses too have an important role to play in a system based on prevention rather than treatment. They find this much easier to do when they can show specialist training and qualifications and when they are able to work independently of the doctor, though under overall medical supervision.

The Royal College of Nursing (RCN) has been responsible for training occupational health nurses. From August 1988, validation of courses and standards became the responsibility of the National Boards for Nursing, Midwifery and Health Visiting. There are three recognised qualifications. The first is the Occupational Health Practice Nurse Award which was obtained after a practical course open to enrolled and registered nurses which demonstrated how nursing skills are adapted for use in an industrial setting. The Occupational Health Nursing Certificate could be obtained after a period of training, either one academic year full-time or two years (six terms) on day release, and covered all aspects of OH. Only registered nurses could become fully qualified OHNs. The third qualification is the Occupational Health Nursing Diploma. None of these qualifications has statutory recognition, so that anyone can work in an OH department without having completed specialist training. As a result, it may be difficult to persuade an employer to allow time off for training. The revolutionary changes in nurse education brought about by Project 2000 will have significant effects on the training of occupational health nurses. Courses are now validated jointly by the National Boards and the universities, and have been upgraded to diploma level. The syllabus has been planned using the framework of the Hanasaari model, developed at a conference in Finland. This stresses the need to regard the total environment in which the workplace is set.

There is no legal obligation on an employer to employ a nurse with any qualification. About half the nurses working in OH do not have any special OH qualification over and above their basic nurse training. Most work full-time and for medium-sized or large employers. Over half are in charge of the occupational health service and 63 per cent work in a situation where a doctor visits the premises only on a part-time basis. More than 10 per cent have no medical practitioner either on the premises or visiting the establishment.

In 1987 the RCN compiled a Code of Professional Practice in Occupational Health Nursing and in 1991 produced a *Guide to Occupational Health Nursing*. The RCN defines the functions of the OH nurse in very much the same terms as the BMA defines the OH physician's role, with the significant addition of the provision of a routine treatment service. In practice, the OH nurse is likely to be spending a significant part of her time in providing and training others in first aid, screening employees, both pre-and post-recruitment, keeping records and generally giving advice. Increasingly, the nurse has been drawn into health education, both about work-related hazards and about general health, like the importance of a good diet and the dangers of smoking.

The Employment Nursing Advisers are full-time employees of the Health and Safety Executive and part of the Employment Medical Advisory Service. ENAs are key figures in occupational health because many nurses are working on their own and have only EMAS to turn to for advice and information. In recent years, much of their time has been spent in advising employers about the provision and training of first-aiders, who, as has been seen, do not have to be qualified nurses.

In 1993 the HSE published Anna Dorward's study *Managers' perceptions of the role and continuing education needs of occupational health nurses*. Predictably, lay managers had a far more limited perception of the role of the occupational health nurse than did doctor managers or nurse managers: they saw them only as providing treatment for illness and injury at work. Only 50 per cent of lay managers supported nurses taking time off to attend courses. It would seem that the occupational health nursing profession still has a need to sell itself to employers as a vital component in health and safety provision. Training initiatives, still under discussion at the time of writing, will have to take into account that the acquisition of more extensive qualifications involves significant sacrifice of time and money for most OH nurses.

1.7 The inter-disciplinary nature of occupational health

The law is on the whole very unspecific about demarcation between doctors and nurses. There is a number of detailed regulations about the supply and administration of medicines. Prescription only medicines (POM) may only be supplied through a registered pharmacy on the written prescription of a doctor or dentist and no person may administer medicines parenterally (administration by breach of the skin or mucous membrane) unless he is either a doctor or a dentist or acting in accordance with the directions of such a person. However, there are exceptions for OH nurses, to be found in statutory regulations made in 1980. These exempt from the above requirements the supply or administration of medicines in the course of an occupational health scheme by a registered or enrolled nurse where the nurse is acting in accordance with the written general instructions of a doctor.

> 'When a doctor signs a general instruction relating to the type of POM a nurse may supply (or in the case of injectables administer), it must be borne in mind that as the doctor is not prescribing for an individual patient, the nurse must exercise professional judgment and discretion when considering which product to administer or supply for the nature of the condition or combination of conditions from which the patient may be suffering.' (RCN: *Occupational Health Nursing Services Handbook* (1991).

Careful records must be kept and the nurse, who should be specifically named in the doctor's authorisation, must never delegate her authority to a first-aider or other nurse (see Chapter 2).

The Misuse of Drugs Regulations 1973 permit registered nurses employed at a place of work to be in possession of controlled drugs (like morphine) for the purpose of administration to persons injured or taken ill at the place of work at which the nurse is employed. The drugs must have been supplied to the nurse by a doctor employed at the place of work for the medical supervision of workers. Security must be efficient and the nurse must keep a register detailing every administration of the drugs. If the drugs are lost or stolen the nurse must report to the police as soon as possible.

Apart from the legislation about medicines, the division between the roles of doctor and nurse in any setting rests essentially on custom and practice. Both professions have something to say about the matter in their ethical rules. The General Medical Council in 1983 ruled that a doctor who delegates treatment or other procedures to a person who is not competent to carry them out is liable to disciplinary proceedings: 'It is also important that the doctor should retain ultimate responsibility for the management of his patients because only the doctor has received the necessary training to undertake this responsibility.' The United Kingdom Central Council, in its Code of Professional Conduct, states that a nurse should acknowledge any limitations of competence and refuse to perform tasks outside those limitations. This is all very well as a general principle but there are many situations in which members of both professions disagree as to what is proper. It is not practically possible to lay down all the tasks which it is acceptable for a nurse to perform. The OH nurse often works in a team with safety officers and hygienists as well as doctors. Often she is on her own when an emergency arises. The UKCC recognises the need for flexibility in its position statement, *The Scope of Professional Practice* (1992): 'In order to bring into proper focus the professional responsibility and consequent accountability of individual practitioners, it is the Council's principles for practice rather than certificates for tasks which should form the basis for adjustments to the scope of practice.'

The RCN has advised its members that the role of the OH nurse is not a static one confined by a list of tasks or duties. The nurse must be conscious of the need to observe the UKCC Code of Conduct, acquire a basic qualification in OH nursing, acquire appropriate skills and attend periodic retraining. However, nurses are warned that the RCN Indemnity Insurance Scheme excludes cover to those undertaking radiography who are not registered members of the Society of Radiographers. Although a joint committee of the BMA and RCN advised that there should be a procedure for consultation between OH personnel to define their relative responsibilities, there can still be friction. The nurses in particular sometimes find themselves in conflict with safety officers. If the problem proves intractable, the best advice is for the nurse to contact her ENA at the local HSE offices.

Doctors and nurses are themselves legally liable for any acts of negligence which cause damage. The OH doctor is only liable for the negligence of the nurse if he was himself negligent in asking her to do something without checking that she was qualified. If they are employees acting in the course of employment, their employer is also vicariously liable for their negligence. It does not follow that, just because an employee is acting in an unauthorised way, he is exceeding the scope of his employment (see Chapter 7).

1.8 The relationship between the occupational health service and the general practitioner

The preventive aspects of the OH specialists need not impinge on the general practitioner's function. If the worker is found to be suffering from some work-related illness, or has an accident at work, the GP should be informed, if the patient gives consent. Treatment is more problematical. The BMA advises that the occupational physician should treat patients only in co-operation with the patient's GP except in an emergency. If he thinks that the worker should consult his GP, he should urge him to do so. The occupational physician should refer a patient directly to hospital only in consultation or in agreement with the GP. He should not influence or appear to influence any worker in his choice of GP. The RCN advises OH nurses not to undertake specific treatments without the request of the GP in writing, other than treatment for minor injuries and in an emergency. Vaccination and immunisation programmes for workers at risk must be supervised by a registered medical practitioner.

An area of potential conflict is the confirmation that absence from work is due to sickness so that the worker can claim sick leave and sick pay from his employer. This is usually the GP's job and he is only obliged by his contract with the Family Health Services Authority to give a certificate free of charge after seven days' absence. However, the employer may ask the OH department for their advice as to the fitness of the worker. The employer's motive may be suspicion that the employee is malingering. Frequent or lengthy absence may indicate a serious health problem which may put other workers at risk. The worker's absence may be causing the employer financial problems and other employees may be showing resentment. The BMA advises consultation with the GP (with the employee's consent). If the GP and the OH doctor disagree, and there is the prospect of disciplinary action, the employer may allow the employee within company procedures the right to go 'on appeal' to an independent consultant (see Chapter 8).

Where the occupational physician is also the worker's general practitioner (a not uncommon event), he must be careful to keep the two functions separate. In such cases, it might be preferable to advise the patient to see another doctor in the practice (if there is one). Information about the employee must not be transferred from the OH to the GP records, or *vice versa*, without the employee's consent.

Chapter 2

The Legal Status and Liability of the OH Professional

2.1 *Servants and independent contractors*

In the medical and nursing professions there is a diversity of outlets for professional skills. In examining the status of the occupational health (OH) doctor and nurse, it is necessary first to analyse the various types of occupational health service which exist today. The main distinction to be made is between the OH service which is set up 'in-house' by the employer, and that which is bought in from an outside consultant who is in business on his own account. Most OH professionals who work full-time for one employer are his 'servants' (employees) in the legal sense, whereas the independent OH consultants are usually what lawyers term 'independent contractors'. A further complication is that, for example, a nurse may be employed by a company like BUPA which provides OH services to employers, but be lent out to BUPA's customers. As regards BUPA she is a servant, but when she visits the company to which she is sent to give advice or examinations she is an independent contractor.

The importance of all this is that the law divides contracts to sell one's labour to another into *contracts of service* and *contracts for services*. A junior hospital doctor or a nurse in an NHS hospital is a servant employed by the health authority or NHS trust under a contract of service. A general practitioner (GP) in the NHS is a self-employed independent contractor employed by the Family Health Services Authority under a contract for services. This is not merely an antiquated conundrum for lawyers: it has vital consequences in practice. The employer is liable vicariously for the negligence of his servant, but not in general of his independent contractor. Many statutory employment protection rights like the right not to be unfairly dismissed are given only to those employed under a contract of service. The employee has PAYE and National Insurance contributions deducted from his wages by his employer; not so the self-employed independent contractor. The self-employed have the benefit of more generous tax allowances under Schedule D rather than Schedule E, but they cannot claim unemployment benefit, statutory redundancy payments, or industrial injuries insurance benefits.

How can an individual determine into which category he falls? The courts have laid down various tests, but none is conclusive. It is instructive to see how the parties themselves have labelled their relationship, but courts have sometimes refused to accept this, as in one case where a manual labourer on a building site was told by the foreman that he was 'on the lump' and would not have tax and National Insurance contributions deducted. When he later had an accident on the site, the court held that he was, after all, an employee and was therefore covered by the Construction (Working Places) Regulations 1966 which did not protect self-employed persons (*Ferguson* v. *Dawson* (1976). In a later case the court, having held that a semi-skilled worker in a factory who had for several years with his agreement been classified as self-employed was in law an employee and could thus complain to an industrial tribunal that he had been unfairly dismissed, sent the papers to the Inland Revenue so that back taxes could be collected (*Young and Woods* v. *West* (1980)). What the law looks for is evidence about the degree of dependence of the employed person. If he is controlled in the performance of his work, is an integral part of the employer's organisation, uses tools and equipment supplied by his employer and cannot provide a substitute to do his job without his employer's permission, he is likely to be an employee under a contract of service. The number of hours worked is not conclusive: many part-time workers are servants rather than independent contractors.

The British Medical Association (BMA) advises that the doctor should be clear whether he is being offered a contract of service or one for services, and that legal advice should be sought on the wording of any draft contract. Strictly speaking, a contract does not need to be in writing to be legally enforceable, but it is easier to prove what has been agreed if there is written evidence. An employer must give a written statement of the basic terms of a contract of employment to his employee within eight weeks of his starting work. This does not apply to independent contractors or Crown servants. It is unlikely that the courts would accept that a full-time doctor or nurse working in-house was anything other than an employee. A nurse working part-time for one company only would probably be in a similar position, but a physician or nurse providing a few hours a week on a sessional basis, especially if he or she visits more than one company, probably falls into the self-employed category. However, in the only case so far to come before the courts, *Westminster City Council* v. *Shah* (1985), the Employment Appeal Tribunal upheld the decision of an industrial tribunal that a GP who acted as a locum occupational health physician for the Council for five morning sessions a week was directly employed by the Council. His normal hours of work were fixed, he was paid a fixed amount for each session, irrespective of the amount of work involved, and all his work was done on the Council's premises as they required. On the other hand he was not subject to their disciplinary procedures, he received no holiday pay and he was assessed (wrongly, in the tribunal's opinion) to income tax under Schedule D. The court emphasised that each case had to be looked at on its facts 'in the round'. 'Certainty in this area of the law seems to be a will o' the wisp in whose pursuit there are always dangers.'

If all that the doctor does is to act as an Appointed Doctor (AD) for the purpose of performing statutory tests he is employed by the Crown, even though he will be paid by the employer of the workers and will have a contract with the employer for the payment of his fees. Doctors and nurses employed by the Health and Safety Executive (HSE) are also Crown servants.

2.2   Liability for criminal acts

The special responsibilities imposed by criminal statutes to promote the health and safety of the workers are most often placed on the shoulders of the employer or the occupier of the workplace, rather than on individual employees. The GP who employs nurses and receptionists is, as an employer, subject to prosecution under section 2 of the Health and Safety at Work Act 1974 (HSWA) if he fails to do that which is reasonably practicable to ensure the health, safety and welfare at work of his employees. Since the abolition of the Crown immunity of health authorities from prosecution under the HSWA, a health authority can be taken to court for a similar failure.

Prosecutions of individuals rather than employing organisations are sometimes brought under section 7 HSWA which imposes a duty on every *employee* while at work to take reasonable care for the health and safety of himself and of other persons who may be affected by his acts or omissions at work and to co-operate with the employer or any other person so far as is necessary to enable him to perform his statutory duties under health and safety legislation. Only if the doctor or nurse is an employee will he be bound by this section. The duty is to take reasonable care, i.e. not to be negligent: the burden of proof of negligence in this section is on the prosecution. A doctor who discovered that a vehicle driver had developed a condition likely to cause him to be a source of danger to other employees or the public might be in breach of this section if, having failed to persuade the driver to tell the employer, he did not inform management or the DVLC in Swansea that the driver was unfit. The duty of confidence yields to the duty to protect others (see Chapter 3).

It is, however, difficult to imagine a case where a health professional who had acted in good faith would be accused of a crime, because the Health and Safety Executive and the local authorities, who alone have power to bring prosecutions, subject to the intervention of the Director of Public Prosecutions, are not wont to prosecute other than in cases of deliberate and flagrant breaches of the law.

2.3   Liability for negligence

In the civil law of tort, a duty of care falls on anyone who is placed in a position where he can as a reasonable individual foresee that his actions may cause harm to others. A tort (called delict in Scots law) is a wrongful act or omission, other than a breach of contract, in respect of which damages can be claimed by the victim from the wrongdoer in respect of loss or injury. The duty is to take the care which the reasonable man would have taken in all the circumstances of the case. Who is the reasonable man? He is the average, ordinary man 'on the Clapham omnibus', as one Victorian judge put it. If he holds himself out as having a particular skill, like a doctor, nurse, solicitor or accountant, he is judged by the standard of the reasonable average member of his branch of the profession; if an action is taken in the courts this standard will be explained by expert witnesses drawn from the same profession.

A failure to take reasonable care will not in itself give rise to a civil action for damages: there must also be damage to a plaintiff or plaintiffs. (The law of negligence will be examined more fully in Chapter 7.) An instructive case was that of *Stokes* v. *Guest, Keen and Nettlefold* (1968). Mr Stokes was frequently required in the course of his employment as a tool-setter by GKN to lean over oily machines. He died of scrotal cancer after fifteen years' employment. The risk of cancer from mineral oil was established by research from the late 1940s. In 1960, the then Factory Inspectorate published a leaflet advising employers to give warnings to and monitor those who worked with mineral oil. GKN employed Dr Lloyd as a full-time OH physician; although he realised the risk, especially as another employee is the same factory had died of scrotal cancer in 1963, he decided that it was too small to make periodic inspections and warnings necessary. Stokes died in 1966 and it was held that if he had been warned earlier he might have survived.

The judge recognised the dilemma of the OH physician:

'A factory doctor when advising his employers on questions of safety precautions is subject to pressures and has to give weight to considerations which do not apply as between a doctor and his patient and is expected to give, and in this case regularly gave, to his employers advice based partly on medical and partly on economic and administrative considerations. For instance he may consider some precaution medically desirable but hesitate to recommend expanding his department to cope with it, having been refused such expansion before; or there may be questions of frightening workers off the job or of interfering with production.'

Dr Lloyd had seen one man in 1962 who had been advised by a specialist and his GP to cease working with oil; he told him to stay at work to keep up his earnings. He had given a talk to the works council in 1963 about the importance of changing working clothes; he had mentioned scrotal cancer in the talk but it had been omitted from the minutes which were generally distributed because they were read by both sexes! It was held that the doctor was negligent in not warning and testing the workers. As he was an employee, GKN were held vicariously liable for his negligence and had to pay damages to Mr Stokes' widow.

Compare *Brown* v. *Rolls-Royce* (1960) in which an employee contracted dermatitis through contact with machine oil. The employers did not supply barrier cream, about which at the time there was a division of opinion in the medical profession. They relied on their medical officer, Dr Collier, and on his advice barrier cream was not one of the precautions required, though most other employers did supply it. It was decided that there was no negligence, because it had not been proved that barrier cream was an effective precaution against dermatitis in this job.

A number of different points arise from these cases:

(1) They demonstrate the inherent dilemma in the position of the

OH professional with duties to both employee and employer, but they also illustrate that the primary duty is to the employee. With respect to the learned judge in *Stokes*, it is not the job of the doctor or nurse to balance the books. Of course, the OH department must exercise commonsense and tact when dealing with both management and trade unions. There will be some suggested changes to the working environment which will make life more pleasant but are not absolutely necessary, and others which are needed to protect against harm. Therein lies the importance of the expertise of the specialist. He must be able to place information about the likely medical risks (based on his reading of Government publications and the results of research published in medical journals) before management so that they can make an informed decision. He should also have some knowledge of possible precautionary measures and their efficacy. If he finds evidence that there is a real risk of injury, it is his duty to communicate it to the employers, however much preventive measures may cost. He cannot thereafter be held responsible if they do not take his advice, though this may eventually be held to be negligence by the managers. In the long run it may prove cheaper to spend money on precautions now than on damages later, quite apart from the human costs.

In a case heard in 1984 James Kellett claimed damages for industrial deafness sustained in the course of his employment in the British Rail Engineering works at Crewe. He was first provided with ear muffs in 1979. His case rested mainly on the argument that he should have been provided with hearing protection at a much earlier date. In a previous decision relating to shipbuilders, *Thompson* v. *Smith's Shiprepairers* (1984)), the court had set the date when the average British employer should have taken precautions against noise-induced hearing loss at about 1963 when a Government report, *Noise and the Worker*, was published. However, in British Rail's employ in the early 1950s was a Divisional Medical Officer at Eastleigh, a Dr Howkins. This exemplary physician wrote to the Chief Medical Officer of BR in 1951 that he had tested samples of ear defenders in the boiler shops and had found them particularly effective. A request from him and from the works committee that the defenders should be provided free of charge was rejected by management because of the cost (1s 3d a pair). In 1955 a proposal from a consultant physician to do research into industrial deafness in the workshops at Swindon was rejected by BR management partly because of fears that it might precipitate claims for damages from the workers. Because of this evidence, the judge in Mr Kellet's case set the date at which BR should have provided ear defenders as 1955, eight years earlier than for industry generally (*Kellet* v. *BR Engineering*). This may not seem so significant until it is appreciated that there were over 2000 claims for industrial deafness made against BR and that the earlier date substantially increased the measure of damages in nearly every case.

(2) They show that the employer is vicariously liable for the negligence of OH professionals as long as they are directly employed. He may also be liable for an independent consultant under the employer's non-delegable duty (Chapter 7), but this is more doubtful. This is one reason why it may be important to ascertain whether there is a contract of service or for services.

(3) They demonstrate how important it is to keep up-to-date, especially with Government publications. Anther case, *Burgess* v. *Thorn Consumer Electronics Ltd* (1983), concerned a Guidance Note from the HSE about tenosynovitis. This was received by Thorn at their factory at Bexhill, but the personnel department failed to recognise it as an occupational hazard for the workers. At the factory, there was no specialist OH assistance; the 'surgery' was staffed by first-aiders. It was held that the employer was negligent in not warning Mrs Burgess that if she started having pains in her wrist or arm she should see her doctor immediately. If she had been warned the condition could have been diagnosed before surgery was needed.

(4) The judge in the Stokes case summarised the duty of the employer (and the same standard will be applied to doctors and nurses) as follows:

'... the overall test is still the conduct of the reasonable and prudent employer, taking positive thought for the safety of his workers in the light of what he knows or ought to know; where there is a recognised and general practice which has been followed for a substantial period in similar circumstances without mishap, he is entitled to follow it, unless in the light of common sense or newer knowledge it is clearly bad; but, where there is developing knowledge, he must keep reasonably abreast of it and not be too slow to apply it; and where he has in fact greater than average knowledge of the risks, he may be thereby obliged to take more than the average or standard precautions. He must weigh up the risk in terms of the likelihood of injury occurring and the potential consequences if it does; and he must balance against this the probable effectiveness of the precautions that can be taken to meet it and the expense and inconvenience they involve. If he is found to have fallen below the standard to be properly expected of a reasonable and prudent employer in these respects, he is negligent.'

Doctors take the attitude that when they examine an applicant for a job or for insurance or entry to a pension fund, they are not in a doctor/patient relationship with that person. They see doctors as having three forms of contact with patients: the traditional therapeutic relationship, that of the impartial medical examiner reporting to a third party, and that the research worker. The implication for some doctors is that they are not as strictly bound by ethical duties in the second and third situations.

'The absence of a patient/physician relationship may result in the absence of an unambiguous duty of the physician to uncover disease, disclose medical data to the patient, advise the employee of risks of further exposure, and protect the confidentiality of the information disclosed and the advice given.' (Samuels: Medical Surveillance, *Journal of Occupational Medicine*, Aug. 1986)

Ethics are for the profession to determine, but it is unlikely that the courts would make the same distinctions. If we take the duty of confidentiality, this is implied by law into all kinds of relationships. It was said in one case that:

'... if the circumstances are such that any reasonable man standing in the shoes of the recipient of the information would have realised that upon reasonable grounds the information was being given to him in confidence, then this should suffice to impose on him the equitable obligation of confidence.'

Surely an individual impliedly confides in a doctor whenever he gives him information in a 'professional' situation?

The duty of care in negligence arises whenever it can be foreseen that a careless act or omission may harm another. There is no duty to assist a stranger, but if the doctor or nurse as it were creates a relationship by examining or testing an individual a duty will arise at least to perform the examination carefully. In Mrs Sutton's case (below) a nurse in a Well Woman Clinic who examined a woman for breast cancer and did not refer her to a doctor when she complained of a lump in her breast was held liable for negligence. A doctor using a healthy volunteer for research is not in a therapeutic relationship but he will be liable if he negligently causes damage to the volunteer.

As will be discussed in Chapter 3, there is no blanket obligation in law to give the patient information about himself, but it is likely that courts would hold a doctor negligent if, having discovered in the course of screening that a worker was showing signs of susceptibility to a substance, he did not at least warn the patient to avoid further exposure. Only if the worker's condition were incurable, could not be treated and could not be passed on to others might there be justification for the doctor's silence. In one incident, a part-time occupational health physician passing through the factory chanced to see a worker stripped to the waist. He noticed swelling of glands and other clearly visible symptoms of Hodgkin's disease. He obtained the man's consent to writing a letter to his GP, who in his turn sent him to a consultant who confirmed the diagnosis and commenced immediate treatment. There is little doubt that the OH physician would have been acting negligently if he had not at least advised the man that he should consult his own doctor. What if a doctor acting as an impartial medical examiner for an insurance company discovers such symptoms? Even though the company expressly forbids him to discuss his report with the applicant, it would be arguably negligent (and, as a mere lawyer, I should have though unethical) for the doctor not to indicate to him that he should seek further medical advice (one possibility is to obtain the applicant's agreement to sending a confidential letter to his GP).

What, then, *is* the difference between the therapeutic and the other relationships? As you would expect, it is the existence or absence of a duty to give treatment. The doctor or nurse who examines an applicant for a job undertakes only to assess his medical suitability. He has no further obligation. He can say: 'I have been asked to examine you by the employer. These are the examinations I propose to do. I shall only perform them with your agreement and I shall only give the results to the employer with your consent. If you refuse consent I cannot write a report about you and in that case I don't think you have much chance of the job, but that is your decision'.

2.4   Liability to the workers

Occupational physicians and nurses have a two-fold duty to the work-force. Since they have no contract with the employees (the contract is with the employer) their duty flows only from the law of tort. In the first place, they assume a broad obligation towards all those employed by the employer while at work. The OH doctor or nurse is in one sense like a general practitioner who has accepted patients on to his list. The list comprises all the workers in that part of the workplace over which he has jurisdiction. He only undertakes to care for the workers while they are at work and he does not usually promise to provide treatment other than in an emergency. Lest this be thought so onerous a duty as to deter any from the practice of occupational medicine, remember that the duty is to take reasonable care, the care that an average professional would take in all the circumstances, taking into account available resources and the degree of risk. A doctor who only visits a factory for one afternoon a week cannot be reasonably expected to monitor the health of the entire workforce in every particular. In medical negligence suits, the ability of the doctor to bring evidence that he is supported by the opinion and practice of other doctors even if there is an opposing school of thought is very good evidence that he has acted with reasonable care (*Maynard* v. *W. Midlands RHA* (1985)).

What if the doctor or nurse becomes aware that an employee is suffering from a disease which he may communicate to other workers or is in some other way a danger – a disturbed patient who is threatening violence, or a driver who is taking addictive drugs? The health professional may have a positive duty to breach confidence for the protection of others and may be liable in negligence if he does not do so. Just as the physician had a duty to warn Mr Stokes about the cancer-inducing mineral oil, he might have a duty to tell him (or at least the personnel department) that the man working next to him in the factory is an unpredictable schizophrenic. The American courts have held a doctor liable for not warning a young woman that one of his patients had murderous intentions towards her, which he had confided to the doctor. After she was murdered by the patient, her parents sued the doctor. The court held that he was negligent in not breaking his patient's confidence (*Tarasoff* v. *University of California* (1976)).

There is much debate at present about AIDS. If a doctor knows that a worker is carrying the AIDS virus is he negligent if he does not reveal it to those with whom the worker comes into contact? The medical evidence is that it is virtually impossible to transmit the virus through normal contact, so that in almost every case it will not constitute negligence if management and fellow-workers are kept in ignorance. Only if there is a real risk of contact with contaminated blood or body fluids which cannot be avoided by standard hygiene procedures employed in every case should the doctor or nurse consider breaching confidence in the interests of others.

Secondly, the OH professional has a duty of care towards any worker who approaches him for advice or assistance. A careless diagnosis, an inappropriate prescription, a failure to refer someone for specialist advice can all constitute negligence. For example, Mrs Sutton went for a health check to a Well Woman Clinic. She told the nurse who examined her, Nurse Hancock, that she thought she had a lump in her breast. The nurse could not feel it, so she did not refer the matter to a doctor as she ought to have done. It was held that the nurse was negligent because she should not have taken upon herself the role of diagnostician. Her employer, a private health organisation, was held vicariously liable despite her disobedience to instructions: she was still acting in the course of her employment as a nurse. The damages were low in this case because the judge found that all that would have been achieved by an earlier referral was a few more years' life (*Sutton* v. *Population Services* (1981)).

The Court of Appeal has ruled that an employer in giving a reference (there is no legal obligation to do so) for an employee or former employee cannot be liable for mere carelessness, because there is a defence of qualified privilege. For legal liability to arise, it must be proved that any false statement in the reference was made from a spiteful motive (*Spring* v. *Guardian Assurance* (1993)). Raymond Petch was employed as a civil servant by the Commissioners of Custom and Excise, later being transferred to the DHSS. He gained rapid promotion, being regarded as a high flyer, until he had a mental breakdown. He claimed injury benefit under the Civil Service Pension Scheme, in essence because the pressures of his job had caused his ill-health. He argued that the stresses of work imposed on his manic depressive personality had been too great to bear. The pension scheme is administered by the Treasury. Their requests for information from his superiors about the reasons for Mr Petch's breakdown led to rejections of his claim, basically because they negligently responded that the work he had to do was not particularly stressful.

Eventually, the court held that Mr Petch would not have suffered the breakdown if he had not been subjected to heavy pressure of work. His persistence was at last rewarded and he received his pension. He then asked for damages for negligence. The Court of Appeal held that the employers owed the plaintiff a duty to take reasonable care to ensure that the duties allocated to him should not damage his health. However, the defendants were not negligent in failing to spot the early signs of his illness and taking steps to prevent it. Nor were they negligent on the facts in overloading him with work, because he was apparently competent and enthusiastic to work under pressure. He was a typical workaholic. With respect, this part of the judgment fails to appreciate that the employer has a duty to try to prevent ill-health in his workers. The consent of the worker to undergo risks does not absolve the employer from his obligation to care for him (see Chapter 7).

The majority of the Court of Appeal decided that although the managers were negligent in relation to their answers to the Treasury's queries, they could not be liable in tort without proof of a spiteful motive (*Petch* v. *Customs and Excise Commissioners* (1993)).

This case raises very important issues with respect to medical reports. Is a doctor or nurse writing such a report immune from liability in negligence to the subject of the report? A report written by one health professional to another *for the purposes of treatment* cannot fall within this principle. A consultant who negligently gives the wrong advice to the patient's GP who in consequence prescribes the wrong drug would be liable to the patient in negligence. Also, a health professional or hygienist who negligently performs screening tests could be held liable to the worker whose early symptoms or over-exposure go unmarked. But occupational health reports are mostly not of this kind. They represent the opinion of the OH professional as to the competence and ability of the worker to do a job, and have financial rather than medical consequences. The occupational physician is asked to write a report on the fitness of a pilot but confuses two employees and attaches the alcoholic's report to the pilot's records. In another example, the doctor negligently advises the employer that a worker is fit to work, and therefore malingering when he takes sick leave, because the doctor has failed to appreciate the worker's true medical condition. In my view, the doctor cannot be held liable to the worker for such errors unless he has a spiteful motive. All the more important to give the employee access to the medical report so that errors do not go unchallenged.

The doctor must also have the employer's interest in mind, because a medical report which is carelessly written in the applicant's favour, resulting in the employer taking on an unreliable employee, might give rise to an action by the employer. For negligence to be established, it would be necessary to show not just that the doctor's assessment of the worker in question was at fault but that there had been a failure to do what a reasonable doctor would have done. An example would be a report of a pre-employment medical examination which stated that an employee with obvious symptoms of a serious heart condition was in the best of health.

Doctors in particular should be warned that it is not merely the content of their writing which needs attention. James Prendergast, an asthmatic, was prescribed Amoxil by his doctor. The writing on the prescription was so illegible that the pharmacist dispensed Daonil. The judge held that both the pharmacist and the doctor were negligent. The doctor had a duty to write clearly and the pharmacist should have checked with him because the dosage prescribed was unusually high for Daonil. The GP was 25 per cent and the pharmacist 75 per cent responsible (*Prendergast* v. *Sam and Dee* (1988)).

2.5 Consent to medical treatment

Every adult is entitled to decide what physical contact he will permit and it is both criminal and tortious to perform any form of medical examination or treatment on him without his consent. The layman describes such an invasion of the person as an assault, but technically an assault is only a threat of physical contact; the unpermitted contact itself is described as a battery. The fact that a patient may need treatment, even life-saving treatment, is irrelevant: the patient must decide. There are several exceptions. Where the patient is unconscious, emergency treatment of a life-saving nature may be performed without consent. It is not legally necessary to obtain the consent of relatives in such a case. Consent may be implied by conduct: the employee who holds out his arm for blood to be taken need not, strictly speaking, be required to sign a consent form. Children of sixteen or over have the right to give consent on their own behalf (Family Law Reform Act 1969) and the parents' views are irrelevant unless the child is physically or mentally handicapped (this is despite the law that full legal independence is only acquired at eighteen). Treatment of young children can only be performed with the consent of at least one parent or the local authority if they are in care; older children under sixteen may give a valid consent if they can appreciate the nature and consequences of the treatment (*Gillick* v. *West Norfolk and Wisbech AHA* (1985)).

There is no battery if the patient knows the broad nature of the proposed treatment and agrees to it. Suppose that an OH physician undertakes a programme of vaccination at the place of work. He obtains the written consent of each individual worker to give injections of a particular vaccine. He commits no battery, but he may still be held liable in negligence if he has not informed the workers before they agreed to participate of any risks inherent in the vaccine to be administered. This is the doctrine of 'informed consent' and it is part of the law of negligence. Doctors and nurses when obtaining the patient's consent must give them the information which a reasonable doctor or nurse would have given that patient (*Sidaway* v. *Board of Governors of the Bethlem Royal and the Maudsley Hospital* (1985)). That case concerned a woman who alleged that she had not been told about an inherent risk of less than 2 per cent of damage to her spinal cord before she consented to have an operation which, though carefully performed, left her partially paralysed. The evidence was that some surgeons would have told, but that others would have kept silent. Mrs Sidaway's action failed.

The doctrine of informed consent does not entitle the patient to all the facts about his case, merely the material facts about the proposed treatment. The English courts have not adopted the American attitude that the patient must be given every detail and have left the issue of what should be communicated to the medical profession. It is they, after all, who alone can identify 'the doctor on the Clapham omnibus'. Note that the signing of a consent form of itself is *never* conclusive in these cases. The law puts less store on the form of the agreement than on the reality of the consent.

The consent of an individual to the taking of blood will prevent it from being either a criminal or civil battery, unless he was misled by false information as to the purpose of the test. A routine blood test for other purposes could be used also to test for HIV antibodies without a battery being committed. As to negligence, however, as there is at present no effective treatment, and the discovery that an individual is infected with the virus can lead to severe consequences, both financial and emotional, it is probably true that a health professional's duty of care to his patient demands that he tests for HIV only after he has obtained his fully informed consent.

The testing of anonymous samples, such that the donor cannot be made aware of the result because the tester does not know his identity, is lawful, whether or not it is ethical. Testing of identified samples without the knowledge of the donor is *not* anonymous; the doctor or nurse can identify the samples. It is the latter kind of procedure which could lead to an action in negligence, because the employee has not given informed consent.

It should be noted that when a statute provides for obligatory medical examination of workers it does not mean that they can be examined without their consent. Such statutes commonly provide that the worker has a statutory duty to present himself for a test, e.g. Reg. 16(7) Ionising Radiations Regulations 1985: 'An employee ... shall when required by his employer and at the cost of the employer, present himself during his working hours for such medical examination and tests as may be required ... and shall furnish the employment medical adviser or appointed doctor with such information concerning his health as the employment medical adviser or doctor may reasonably require'. If the employee refuses, the doctor must respect that decision although the employer will then be justified in taking disciplinary action.

2.6   Liability to the employer

The OH professional will have a contract with the organisation which employs him, so he will have duties both in contract and in tort. If he is employed under a contract of employment, the law implies many obligations into the relationship which form part of the agreement even if nothing has been said or written. The duty of trust and fidelity discussed in Chapter 3 is a good example of an implied term. One other significant duty is to take care in carrying out his job. In a case in 1956 one employee negligently injured another by reversing a vehicle into him on the employer's premises. The employer paid compensation to the injured employee and claimed on his employer's liability policy. Insurance companies have a right of subrogation, i.e. they take over all the rights which the insured person had when they pay out on the policy. Acting in the name of the employer, the insurance company sued the careless driver for all the loss which he had caused his employer by his negligence. The House of Lords decided that negligent employees have a duty to compensate their employer for the damages he has been forced to pay because of their carelessness (*Lister* v. *Romford Ice and Cold Storage*)). After establishing the point of principle, the insurance companies indicated that they would not in future reimburse themselves by pursuing employees, unless there was wilful misconduct or collusion. Also, in the later case of *Morris* v. *Ford Motor Co.* (1973), the Court of Appeal refused to allow an insurance company to sue the driver of a fork-life truck who had negligently injured the employee of a subcontractor.

Thus, where an employer is insured against liability to his employees caused by the negligence of directly employed OH personnel (as is required by the Employers' Liability (Compulsory

Insurance) Act 1969), it is unlikely that an indemnity will be obtained by the employer's insurance company from the doctor or nurse's protection society or professional indemnity policy. The same is now true of OH personnel working in the National Health Service ('Crown indemnity').

2.7   Liability to the public

The health professional does not have a *legal* duty to provide medical care for those whom he has not accepted as patients. There is no duty in English law to act as a Good Samaritan and go to the aid of someone lying seriously injured in the street, but if you do you will have an obligation to take reasonable care. What if an employee of a subcontractor, or a visiting member of the public, is taken ill or has an accident on the premises where the OH doctor or nurse is at work? The BMA in *The Occupational Physician* (1984) advises that the OH doctor takes full medical responsibility only for those *working* on site, including contractors, but to the extent that he is concerned with the effect of work on health, he is advised that included therein is the health of the public at large, either in general or as individuals.

It is unlikely that a health professional would be held to owe a *legal* duty to the general public, other than in a case where he negligently permitted a worker known to him to be 'dangerous' to be at large, without giving any warning to the proper authorities. His obligation to the wider community is a moral and ethical duty. In an extreme case, it might oblige him, against the wishes of his employer, to reveal dangers to the public of which they are in ignorance (see Chapter 3). Mostly, of course, the OH professional is unable to do more than advise and warn, but should do whatever is reasonable in all the circumstances of the case.

2.8   HIV-infected health care workers

Ill-informed fears about the dangers of contracting AIDS from an HIV-infected health professional, fuelled by worldwide publicity about the Florida dentist who allegedly infected several of his patients, have led to the setting-up by the Department of Health of the UK Advisory Panel for health care workers infected by blood-borne viruses. This has a strong occupational health representation. The Expert Advisory Group on AIDS (EAGA) in 1994 produced guidelines for health authorities and health professionals on the management of HIV-infected health care workers. These provide, *inter alia*:

(1) All health care workers should scrupulously adopt safe working practices to prevent transmission of HIV infection.
(2) HIV-infected health care workers should not perform exposure-prone procedures. (These are procedures where there is a risk that injury to the worker may result in exposure of the patient's open tissues to the blood of the worker. They include those where the worker's gloved hands may be in contact with sharp instruments, needle tips or sharp tissues (spicules of bone or teeth) inside a patient's open body cavity, wound or confined anatomical space where the hands or fingertips may not be completely visible at all times.)

   Where there is doubt, expert advice should be obtained from a specialist occupational physician who may in turn wish to consult the Panel.
(3) Health care workers have an ethical duty to patients. Those who believe that they may have been exposed to HIV infection in their personal life or in the course of their work must seek appropriate expert medical advice, and, if appropriate, diagnostic HIV antibody testing.
(4) Infected health care workers who have performed exposure-prone procedures must cease these activities immediately and seek expert occupational health advice. A nominated medical officer in the employing authority should be notified on a strictly confidential basis. If a 'look-back' study (notice to the public together with help-lines and the offer of HIV tests for any patient concerned about possible infection) is being contemplated, the local Director of Public Health must be consulted.
(5) Physicians who are aware that infected health care workers under their care have not sought or followed advice, and are continuing to perform exposure-prone procedures, have a duty to inform the appropriate professional body and the nominated medical officer.
(6) Employers must make every effort to arrange suitable alternative employment, or early retirement, for HIV-infected health care workers.
(7) All patients who have undergone an exposure prone procedure where the infected health care worker was the sole or main operator should, as far as is practicable, be notified of this (in a look-back study).
(8) Employers have a duty to keep information about an HIV-

infected worker confidential, unless the individual consents to disclosure, or, in exceptional circumstances, where it is considered that patients need to be told for the purpose of treatment or the prevention of the spread of infection. This duty does not end with the death of the worker (though if AIDS is mentioned on the death certificate, this is a public document to which all have access).

A failure of a professional who knows or suspects that he is HIV-positive to seek medical advice and act on it is regarded as serious professional misconduct by the medical, nursing and dental professions. The UK health departments published in 1993 a booklet on practical guidance for health authorities on look-back studies. An epidemiological overview of the risk of transmission of HIV from health care worker to patient has been commissioned. Thus far, the evidence is that the risk is tiny and that it is more likely that the health professional will contract HIV from the patient (148 cases worldwide by December 1992).

The Government is at present opposed to the introduction of compulsory testing either of health care workers or patients. The test is unreliable in the first months of the infection, and at best can provide only a 'snapshot' of the individual's HIV status on the day of the test.

2.9   HBV-infected health care workers

Hepatitis B is also a virus, but it presents different problems from the HIV virus. There is a reliable vaccine, and medical treatment results in most cases having a happy outcome, though it is still a serious disease. There is little social stigma attached to infection, unlike HIV infection. Nevertheless, it is far more likely that hepatitis B will be transferred in the course of medical, dental or nursing practice. The Department of Health published recommendations in 1993 and hepatitis B has been added to the remit of the UK panel of experts.

The key recommendations are:

(1) All health care workers should follow general infection control guidelines and adopt safe working practices.
(2) All health care workers who perform exposure-prone procedures, and all medical, dental, nursing and midwifery students should be immunised against hepatitis B, unless immunity has been documented. Their response to the vaccine should subsequently be checked.

(3)   Health care workers who are hepatitis B e antigen (HBeAg) positive should not perform exposure-prone procedures in which injury to the worker could result in blood contaminating the patient's tissues.

(4)   Health care workers who are hepatitis B surface antigen (HBsAg) positive but who are not HBeAg positive need not be barred from any area of work unless they have been associated with the transmission of hepatitis B to patients whilst HBeAg negative.

(5)   Staff whose work involves exposure-prone procedures and who fail to respond to the vaccine should be permitted to continue in their work provided they are not e antigen positive carriers of the virus. Inoculation incidents must be treated and followed up.

(6)   Health Authorities and Trusts should ensure that members of staff employed or taking up employment or other health care workers contracted to provide a service which involves carrying out exposure-prone procedures are immunised against the hepatitis B virus, that their antibody response is checked and that carriers of the virus who are HBeAg positive do not undertake such procedures.

(7)   Occupational health departments should be involved in developing local procedures for managing HBV-infected health care workers.

(8)   Employers should make every effort to provide alternative employment should this be needed.

(9)   The UK Advisory Panel is available when specific occupational advice cannot be obtained locally.

Hepatitis B is a prescribed industrial disease for health care workers. Benefits are also available under the NHS Injury Benefits Scheme for NHS staff.

2.10   Professional indemnity

Those who work for others rather than themselves often assume that any legal liability falls on the employer alone. This is definitely not the law. The primary liability for negligence lies with the person who commits the negligent act. Sometimes there is no alternative defendant. An independent consultant is the only possible defendant if he makes a careless mistake. If he is in partnership, his partners are also liable. Where the doctor or nurse is an employee, the employer is vicariously liable: they are both responsible. If the employer is in financial difficulty and uninsured, or claims a contribution or indemnity from the employee, or where the plaintiff out of revenge chooses to pursue the employee instead of or as well as the employer, damages and costs may threaten. The OH professional may be involved in a HSWA prosecution in which he may need legal advice and representation (it is not possible to insure against a criminal fine). For these reasons, every professional should be covered by professional liability insurance, despite the lack of a legal obligation to insure. Nursing staff commonly insure through membership of the Royal College of Nursing (RCN) or a trade union, doctors have professional indemnity through the Medical Defence Union or the Medical Protection Society. The latter are not, strictly speaking, insurance companies, because they have a discretion whether to indemnify (*Medical Defence Union* v. *Department of Trade* (1979)).

Nurses are often troubled that they are not covered if they undertake 'extended role' tasks without proper authority. Because essentially the proper role of a nurse rests on custom and practice, it is impossible to lay down a definitive list of nursing tasks. The RCN in its guidance notes for the occupational health nurse advises that each OH department should draw up an agreement about the tasks which the nurse is expected to perform. No nurse should undertake any procedure which she is not trained and competent to perform. The OH nurse is not covered by RCN indemnity insurance when she undertakes radiography. The RCN is reluctant to draw up lists of duties which could be regarded as a restriction on the development of the OH nurse's role in a true professional sense. On the other hand, OH nurses need to recognise those areas of expertise which rightly belong to other professions, such as medicine, radiography or physiotherapy. The emphasis is on demarcation agreements, especially with doctors who carry the ultimate clinical responsibility, and on training to enable the nurse to extend her role competently.

2.11 The conduct of research

Research on humans is of two kinds: clinical research on ill patients and research on healthy volunteers. In law, both need the informed consent of the subjects. As has been seen, this allows a doctor or nurse to withhold information if a reasonable doctor or nurse would have kept silent, as long as the subject knew the broad nature of what was to be done.

Ethical rules are laid down in the Declaration of Helsinki. The Royal College of Physicians produced a second edition of its *Guidelines on the Practice of Ethics Committees* in 1989. The Department of Health produced new guidelines for research ethics committees in 1991. The Helsinki Declaration demands that every subject '... must be adequately informed of the aims, methods, anticipated benefits and the potential hazards of the study and the discomfort it may entail ... the doctor should obtain the subject's freely given informed consent, preferably in writing'. It makes an exception for clinical research if the physician considers it essential not to obtain informed consent and has discussed the matter with an ethics committee, but this is not recognised in English law. The Department of Health draft guidelines originally incorporated an exception similar to that in the Helsinki Declaration, but this was omitted from the final version.

NHS employees who fail to obtain ethics committee approval for a research project would be subject to disciplinary action. The professional bodies would be likely to regard a failure to obtain ethical approval as professional misconduct.

The College of Physicians in 1986 made the following recommendations relating to research on healthy volunteers:

(1)   All research involving healthy volunteers should be approved by an ethics committee.
(2)   All studies should be scientifically and ethically justified.
(3)   Confidentiality should be maintained (see Chapter 3).
(4)   No study on healthy volunteers should involve more than minimal risk and there should be full disclosure of risks.
(5)   There should be no financial inducement or any coercion that might persuade a volunteer to take part in a study against his better judgment. Payment should be related to expenses, inconvenience and discomfort, not risk.
(6)   The volunteer should be asked to give permission to the researcher for his general practitioner and, if appropriate, 'a company or other medical officer' to be contacted for details of past history. Where appropriate, he should be medically examined and be asked about relevant medical history. He should sign a consent form.
(7)   Any significant untoward event occurring during or after a study affecting a volunteer should be communicated to the general practitioner and appropriate medical action taken to safeguard the volunteer's health.
(8)   The sponsor, whether this be a commercial organisation,

university, NHS or other institution, should agree to pay compensation for any injury caused by participation in a research study without regard to proof of negligence.

Research on healthy volunteers often takes the form of a drug study. The Association of the British Pharmaceutical Industry (ABPI) in 1988 published its *Guidelines for Medical Experiments in Non-Patient Human Volunteers*. Its recommendations are similar to those in the Royal College of Physicians reports. Where research is proposed on a 'captive audience', e.g. medical students or employees, no-one should be made to feel under obligation to volunteer, nor should they be disadvantaged in any way by not volunteering. Volunteers may be rewarded in cash or in kind, but the amount should be reasonable and related to time, inconvenience and discomfort, not risk. The ABPI operates an *ex gratia* scheme whereby any healthy volunteer in a drug trial mounted by an ABPI member will receive compensation for any injury arising from the trial.

In an occupational health setting, it will also be necessary to obtain the agreement and support of the employer, and probably the trade unions. The OH physician must be very careful to ensure that any volunteers have freely consented to take part. He should make clear at the start his role as a research investigator. 'Occupational physicians may be perceived by workers to be part of management and therefore it is particularly important to ensure that informed agreement is given freely and that individuals recognise that they are free to withdraw at any time without detriment. Consent of trades unions to participate in research projects should not be taken to imply the consent of all individuals involved.' (*Faculty of Occupational Medicine Guidance on Ethics* (1993)).

Drugs research using patients is regulated by the Medicines Act 1968. The Committee on Safety of Medicines must license human trials by granting a clinical trial certificate (or exemption from the need to hold such a certificate) and will only do so if satisfied with preliminary research and animal tests. Any adverse reactions must be reported to the committee. This does not apply to non-therapeutic research because it is not regarded as the administration of a medicinal product. Contraception is defined as therapeutic.

Epidemiological research using data only is discussed in Chapter 3.

2.12 The rights of OH professionals

Thus far we have examined only those obligations which the law places on doctors and nurses working in industry. They will be relieved to read that they also have rights. The relationship with the employer will, as has been explained, rest on a contract. It is advisable to insist on a written agreement, and employees (but not the self-employed) can demand a written statement of pay, hours, holidays, pension and sick pay provision, periods of notice and disciplinary rules and grievance procedures, under the Employment Protection (Consolidation) Act 1978 (EPCA). This statement must contain a job title, but not necessarily a job description. If the employee's duties are not spelled out and a dispute arises, there may have to be reference to the job advertisement, what was said at the interview and custom and practice.

The Trade Union Reform and Employment Rights Act 1993 gives the right to complain to an industrial tribunal and to claim compensation from the employer to any employee who has been designated to carry out activities in connection with preventing or reducing risks to the health and safety of employees at work and who has been disciplined or dismissed by the employer for carrying out or proposing to carry out such activities (see Chapter 8).

An interesting case was that of *Woodroffe* v. *British Gas* (1985). Miss Woodroffe was a State Registered Nurse who in 1980 entered the employment of British Gas as an OH nurse. She was given a job description which did not make any express reference to taking blood samples or giving talks on health matters to other employees. Shortly after she took up her post, a new Medical Officer was appointed. He thought that she should take blood samples and give health education talks. Miss Woodroffe explained that these tasks were classified at the time as 'extended role' and that she had not been trained to carry them out. The doctor offered her training which she refused. After a while, he persuaded her to take on these extra responsibilities, which she did with reluctance. The Medical Officer became dissatisfied with her work, including her keeping of records. He instituted an enquiry which he presided over and which recommended her dismissal. She appealed to a higher level of management, and was represented by a member of her professional body, but the appeal was dismissed. A complaint to an industrial tribunal of unfair dismissal, and subsequent appeals to the Employment Appeal Tribunal and the Court of Appeal were also unsuccessful.

Nurses will be interested to note that it was the opinion of the Employment Appeal Tribunal that if Nurse Woodroffe had refused to carry out the additional tasks altogether she could not have been criticised, for they were not tasks which she had been hired to perform. If she had been persuaded to receive specialist training in occupational health the difficulties might have been resolved. There is no legal duty on the employer to give money and time off for further training, so if the doctor or nurse wants this, he or she should negotiate it as part of the contract.

The employer's duty to take reasonable care for the employee's health and safety applies equally to the occupational health department. The employee is entitled to refuse to put himself in a position of imminent personal danger, as did the bank employee who declined to work in Turkey where he had previously been sentenced to death (*Ottoman Bank* v. *Chakarian* (1930)). Doctors and nurses frequently disregard their own safety to go to the assistance of their patients. One such was Dr Baker. Contractors were cleaning a deep well. They installed a petrol pump on a ledge 29 feet below ground, creating dangerous fumes in a closed space. Two workers were overcome by fumes and were lying down the well unconscious. Dr Baker was urged not to go down but he said: 'There are two men down there. I must see what I can do for them.' He died on the way to hospital. When his widow sought compensation from the contractors, they argued that the doctor had caused his own death by his foolhardiness. The widow submitted that it was reasonable for a doctor to take risks to try to save the lives of workers who had been placed in mortal danger by their employer's lack of care. The Court of Appeal held the defendants liable. 'Bearing in mind that danger invites rescue, the court should not be astute to accept criticism of the rescuer's conduct from the wrong-doer who created the danger' (*Baker* v. *Hopkins* (1959)).

Chapter 3

Medical Records and Confidentiality

3.1 The ownership of records

Many of the legal problems of occupational health (OH) workers involve the maintenance, disclosure, transfer or destruction of records of tests and clinical examinations. Central to all these issues is the question of ownership and the right to control medical records. Any words which are reduced to writing comprise two elements: the physical paper on which the words are written and the information contained therein. In the National Health Service, the medical records of patients clearly belong physically to the Family Health Services Authorities (in the case of general practitioner (GP) records) and health authorities, including NHS Trusts (in the case of hospital records) and ultimately, therefore, to the Secretary of State. This is because the contracts of health service personnel provide that patients' records must be made on health service forms. It is also, practically speaking, the most sensible system, because doctors come and go and records must be kept by a permanent central organisation. Only the records of private patients belong to the health professional with whom they have a contract.

Do the same rules apply to the OH records of *employees* of the health authorities? It is probable that these, like all occupational health records, both within and outside the NHS, are physically the property of the employer, because they will be recorded on the employer's paper and held on the employer's premises. In addition, if the doctor or nurse is directly employed under a contract of employment, there is an implied term in the contract that the fruits of the employee produced in the course of employment (and this will extend to patent rights to inventions, and copyrights in written publications, like research findings) belong to the employer.

Doctors and nurses need not feel aggrieved that this is the law, for it does not necessarily follow that the owner of the records has

61

the right of control over the information contained therein. The general practitioner is in control of a patient's records while he continues to have that patient on his list. There is nothing to prevent the occupational health department having an agreement with the employer whereby the health professionals and their staff have exclusive access to medical records. This arrangement, to be watertight, should be incorporated in a written agreement which can form part of the doctor's or nurse's contract, just as many companies have standard contracts with employees about the ownership of inventions.

The right to control medical records will carry with it increased responsibilities. First and foremost, there is the duty to take care of the records and ensure that they are only read by authorised personnel. In the days when documents were held in filing cabinets, it was important to keep them locked; nowadays, more sophisticated controls are necessary to keep computer-held data from being made public, especially where a system is shared with other users or there is telephone access. Passwords may have to be built in and security procedures instituted. In particular, visual display units should not be sited where casual passers-by can read them. Where data are held on computer the controller of the data will have duties under the Data Protection Act 1984.

The physician may move to another job or may retire. The professionally qualified nurse may be made redundant and replaced only by unqualified first-aiders. If the medical records are destroyed this could harm the workers, both individually and as research data which could benefit all. As the records belong to the employer they should be left where they are, but the control be handed over to the new doctor or nurse. What if there will be no new health professional? In a case where a business is closing down, the Faculty of Occupational Medicine recommends that if an organisation is to be closed and will cease to exist, it will be necessary to ensure that records are transferred with individual consent either to each employee's own doctor or to another medical adviser. Similar advice would presumably be given in a case of closure of an OH department. The doctor or nurse might also consider giving a copy of the results of periodic examinations to the worker himself, failing a national system of such records as exists for ionising radiation workers (National Radiological Protection Board), or they may be offered to the Health and Safety Executive (HSE). The Faculty recommends that in the last resort the physician should destroy his records, other than those where retention is required by statute, but this could be risky advice if, as has been argued, the records are not his property.

3.2   How long should records be kept?

The obvious answer is only as long as is necessary, but how long is that? Occupational diseases may take twenty or more years to develop. Patients died of mesothelioma decades after their only contact with asbestos while making gas masks during the Second World War. Occasionally, there is a statutory rule. The Ionising Regulations 1985 oblige the employer to preserve records of tests on employees exposed to radiation for 50 years. The Control of Substances Hazardous to Health (COSHH) Regulations provide that the health record of health surveillance procedures carried out on employees exposed to hazardous substances shall be kept for at least 40 years from the last entry. Otherwise, common sense must be used. The DH recommends in a circular to health authorities that personal health records (other than records held by Family Health Services Authorities) should be held for a minimum of eight years after the conclusion of treatment. Obstetric records should be kept for at least 25 years, and those relating to mentally disordered persons for at least 20 years. The British Medical Association (BMA) recommends a minimum of ten years after the employee left the employment, but advises that records of significant episodes, exposures or accidents should be preserved beyond that limit. However, it must not be forgotten that the employer has an interest in the preservation of records which he may later need to defend an allegation of negligence. The best course is probably to discuss the matter with the employing organisation who will be likely to agree to provide extra storage facilities for records which it wants to preserve.

3.3   The duty of confidence

The duty to keep secrets is enshrined in all codes of medical ethics. All health professionals know that it is one of the strictest ethical rules of their various professions that they should not give confidential information about a patient to third parties other than in exceptional cases. If this obligation is breached, a doctor or nurse may find himself before the professional conduct committee of his

professional body (the General Medical Council (GMC) for doctors
and the United Kingdom Central Council (UKCC) for nurses,
midwives and health visitors) and may even run the risk of being
struck off the register, depending on the seriousness of the offence.
The law leaves the matter of professional ethics to the professions
themselves, so it is from them that guidance must be obtained on
what constitutes good professional practice.

There is a legal as well as an ethical duty of confidence. Lawyers
sometimes find it difficult to define the legal category into which
the duty of confidence falls. For our purposes it is sufficient to state
that the duty arises where there is a contract between the confider
and the confidant which includes a term (which may be express or
implied) that confidence will be kept, or there is a relationship
between the parties which implies a duty of confidence.

An example of the first is a contract of employment, and of the
second the marital relationship. There is little doubt that a health
professional has such a duty, either through contract (doctor and
private patient) or through the general duty of fidelity (occupational health physician and patient, NHS practitioner and
patient). Therefore, if an OH physician were to reveal to the personnel department the fact that an employee had a drink problem
so that she was not considered for promotion, the physician could
probably be sued for damages for financial loss. Where the patient
suffers only mental distress there is probably no right to damages,
but the Law Commission has recommended that damages should
be obtainable. Are there any circumstances in which the physician
is entitled to break confidence?

The GMC lists a number of exceptions to the duty of confidence.
These are as follows:

(1) The patient or his lawyer gives written consent.
(2) The doctor shares information with other health professionals
     concerned with the care of the patient for the purposes of that
     care.
(3) If it is undesirable on medical grounds to speak directly to the
     patient, it may sometimes be permissible to discuss the patient
     with a close relative, or, very exceptionally, a third party other
     than a relative.
(4) Where Parliament requires disclosure, for example of noti-
     fiable diseases.
(5) Where a court of law orders the doctor to disclose information.
(6) Disclosure in the public interest, e.g. to the police about a
     serious crime.

(7) For the purposes of medical research approved by an ethics committee.

The Code of Professional Conduct of the UKCC states that a nurse should 'respect confidential information obtained in the course of professional practice and refrain from disclosing such information without the consent of the patient/client, or a person entitled to act on his/her behalf, except where disclosure is required by law or by the order of a court or is necessary in the public interest'. A further UKCC Advisory Paper on Confidentiality elaborates this obligation. It sets out the following summary of the principles on which nurses should base their professional judgment:

(1) That a patient/client has a right to expect that information given in confidence will be used only for the purpose for which it was given and will not be released to others without their consent.
(2) That practitioners recognise the fundamental right of their patients/clients to have information about them held in secure and private storage.
(3) That, where it is deemed appropriate to share information obtained in the course of professional practice with other health or social work practitioners, the practitioner who obtained the information must ensure, as far as is reasonable, before its release that it is being imparted in strict professional confidence and for a specific purpose.
(4) That the responsibility either to disclose or withhold confidential information in the public interest lies with the individual practitioner, that he/she cannot delegate the decision, and that he/she cannot be required by a superior to disclose or withhold information against his/her will.
(5) That a practitioner who chooses to breach the basic principle of confidentiality in the belief that it is necessary in the public interest must have considered the matter sufficiently to justify that decision.
(6) That deliberate breaches of confidentiality other than with the consent of the patient/client should be exceptional.

Note that the UKCC envisages that confidential information will need to be shared with social workers involved with the care of the patient, as well as health professionals. The UKCC Advisory Paper also appreciates that access to records will be available to persons without any professional code of ethics, like secretarial staff. The

1984 Körner Report on the protection and maintenance of confidentiality of patient and employee data in the NHS considered it essential for social workers who are part of the health care team to have access to clinical data. The code of ethics of the British Association of Social Workers imposes a duty of confidence and some health authorities have developed joint policies. Secretarial staff who have no code of ethics should be trained in the importance of confidentiality and warned that their jobs are at risk if they wrongly reveal information. In the context of occupational health, it is especially vital to keep clinical records separate from other personnel records. No employee who consults the OH department should be assumed to have impliedly consented to the involvement of anyone outside it, even the welfare officer. Clerical staff must have it explained that they are responsible only to the OH staff for records, which should not normally leave the department.

The legal justification for involving other members of the health care team and clerical and social workers is probably that the patient has impliedly consented to disclosure in the normal course of patient care. If this is correct, it means that the patient can withdraw his consent by, for example, ordering the doctor or nurse not to discuss his case with another health professional or social worker. Of course, if the case is one where the 'public interest' is involved, like one of non-accidental injury, disclosure may be justified for that reason. What of the patient who refuses to talk unless he is promised that no record will be kept and no-one else involved? The professional will have to exercise judgment. Imagine an occupational health nurse who, having been sworn to secrecy, is then told that the employee is taking illegal drugs which are affecting her in such a way that she may injure herself and others. The nurse would probably be legally and ethically entitled to report the matter at least to the doctor, if there is one, or, if not, to the employer, but her credibility with the other workers might be seriously damaged. If she does not pass on the information or make a record, later treatment by another acting in ignorance may be prejudiced, and there may be a serious accident. It may be better not in advance to accede to a patient's request that confidence be complete, but to warn that in exceptional circumstances information might have to be passed on.

The Faculty of Occupational Medicine of the Royal College of Physicians in its *Guidance on Ethics for Occupational Physicians* states that access to clinical data should be confined to the OH physician and the OH nurse, although clerical support staff and other members of the OH department may also need to see it. 'Normally, the written informed consent of the individual is required before access to clinical data may be granted to others, whoever they may be and whether professionally qualified or not, e.g. solicitors, insurers, managers, trades union representatives, employment medical advisers, etc.' It reminds OH physicians that, with regard to their own colleagues and staff, when they are acting as line managers they should not demand any wider access than would be given to other managers.

One situation which regularly causes conflict between the OH and personnel departments is that in which the employer is threatened with legal action by the employee, either for compensation for a work-related injury or for unfair dismissal. If the OH department is known to give information too readily, its standing with the workforce will be irreparably damaged. Judicial guidance on this matter is to be found in the decision of the Court of Appeal in *Dunn* v. *British Coal Corporation* (1993). The employee was a 53-year-old miner who sustained injuries to his neck while working at the coal face, as a result of which he was unable to work as a miner. He sued his employer for damages. His solicitor obtained a report from a consultant orthopaedic surgeon which stated that Dunn had no prior history of neck pain. The employers doubted this medical evidence. They wished to have Dunn medically examined by their own expert, and asked that he should be given access to the plaintiff's hospital records, his GP records and the employer's own occupational physician's notes.

> 'So far as the latter is concerned, the employer's solicitors took the view, correctly in my opinion, that although these records had been compiled by their own servant or agent the employee was entitled to claim that as between him and the doctor they were confidential.' (Stuart-Smith L J)

The employee refused consent to disclosure of any records other than those directly related to the accident. The employers asked the court to order the disclosure of all the medical records, because they suspected that the plaintiff at the time of the accident was suffering from a pre-existing medical condition which might have forced him to give up work in any event. This would, of course, substantially reduce any damages he was awarded for loss of earnings.

The court ordered that the action could not proceed until the employee gave consent to the disclosure of all his medical records (including his occupational health records) to his employer's medical expert. The documents would be received by him in confidence and he would have to respect that confidence, except in so far as it was necessary to refer to matters which were relevant to the litigation.

'Thus, for example, the fact that an employee was or had been suffering from some sexually transmitted disease would be quite irrelevant unless it affected his future earning capacity, as might be the case if he was suffering from AIDS.'

The OH records of an employee who is bringing a legal action against his employer, whether in the court or the industrial tribunal, should not, therefore, be disclosed to the employer without the consent of the employee or a court order.

3.4 The consent of the patient

This may, strictly speaking, be given orally, but for the protection of the doctor and the nurse it should always, if possible, be obtained in writing. Where a patient has authorised disclosure to a third party, he may not appreciate that his full medical records over a number of years may be produced, so that the OH physician or nurse should clarify with him the extent to which he is willing to authorise disclosure. However, it would be wrong for a physician to allow the patient to offer misleading evidence by presenting an incomplete picture. The employee who is claiming damages for noise-induced hearing loss and who authorises the doctor to disclose to his solicitors the results of audiograms, while asking him to suppress notes of conversations with the patient about the dangers of his part-time job in a disco, should be told that all matters relevant to deafness must be included, or, alternatively, that no disclosure will be made.

In the occupational setting, the issue will often be one of advising the employer or the trustees of the pension fund whether the employee is fit for work. 'Advice given to management about the results of a medical assessment should generally be confined to advice on ability and limitations of function. Clinical details should be excluded and even when the individual has himself given clinical information to management, the occupational physician should exercise caution before confirming any of it. If a report on the health and fitness for work of an employee is to be communicated to management it is important to ensure that the employee understands the physician's duty to the employer. The contents of the proposed report should be discussed with the employee and consideration should be given to obtaining the signed consent of the employee to be examined in such circumstances.' (Faculty of Occupational Medicine: *Guidance on Ethics* (1993).)

3.5   Relationship with other health professionals

It is a breach of confidence to disclose facts about the patient even to another doctor or nurse, so that if such disclosure is to be made, it must be justified as in the patient's or the public interest or by showing that the patient either expressly or impliedly agreed. Where a patient is in the care of a team of health professionals, it is fairly easy to imply that he is willing to allow all members of the team access to his records. Communication with the patient's general practitioner, or to another occupational physician employed by a different organisation (which may include a company in the same corporate structure but legally separate) should normally be only with his consent, unless there is some special reason why the information should be transferred in the patient's or the public interest. This means that if the employee moves to a different employment he should be asked to give written consent to the transfer of medical records to his new employer. The need for care in such matters is illustrated by one case where an occupational physician routinely notified an employee's GP of the employee's raised blood pressure. Later, the man failed to obtain life assurance, because the insurance company insisted on a medical report from the GP.

Exceptionally, it may be permissible in the patient's interest to communicate with his family or another doctor, for example, if the employee is displaying suicidal tendencies at work, is thought to be a real danger to himself and cannot be persuaded to seek help. This may be justified by the argument of necessity.

3.6   Disclosure of information to researchers and auditors

It is important for records to be kept of accidents and diseases of workers because the emergence of a pattern can give early warning of the necessity for preventive action. The Declaration of Helsinki, the international ethical code on research on humans, lays down guidelines on the conduct of research which have already been discussed in Chapter 2. Epidemiological research may not necessitate the performance of additional procedures, but merely the collection and analysis of data. The OH professional who does his own research will be in control of his own records, but he may be involved in a project in which others ask for access to identifiable patient data. This may also occur during the process of medical audit. Any research protocol must be well-designed and it should be discussed with an ethics committee before commencement; if one is not available, the researcher may approach the Faculty of Occupational Medicine. Remembering that the doctor and nurse are being paid by the employer, permission should also be sought from management. Good industrial relations practice dictates consultation with the trade unions.

The General Medical Council states that medical teaching, research and medical audit necessarily involve the disclosure of information about individuals, often in the form of medical records. 'Where the disclosure would enable one or more individuals to be identified, the patients concerned, or those who may properly give permission on their behalf, must wherever possible be made aware of that possibility and be advised that it is open to them, at any stage, to withhold their consent to disclosure' (*Professional Conduct and Discipline: Fitness to Practise* (1993)).

'Mortality studies have made invaluable contributions to occupational medicine ... One problem concerns the tracing of those who have left their workplace for any reason, to find out whether they are alive or dead. Although death certificates are public documents, the name, age and last known address must be passed to the researchers in order to trace such people. It is considered that provided this information is passed in strict confidence, to be used only for purposes of establishing the fact of life or death, and provided that any final report contains no personal data, this procedure is acceptable' (Faculty of Occupational Medicine: *Guidance on Ethics* (1993)). The Faculty advises that investigation into cancer which involves the obtaining of information from the National Cancer Registry should be approved by a local research ethics committee.

3.7 *Disclosure in the public interest*

'The public interest' is a phrase of venerable antiquity which is employed by the courts to excuse invasions of private rights. In English law it is invoked where the security of the State is at risk, or where the welfare of a community, or a part of it, is in jeopardy. Most of the leading cases concern the Government's attempts to gag former employees, or involve commercial concerns who try to protect trade secrets. Who defines the public interest? In the end it must be the judges themselves, and their views will change and develop. For many years the courts have applied a doctrine that 'there is no confidence in an iniquity'. If the patient or client has been guilty of a crime, he cannot seek the protection of the law to help to keep it secret. The patient who confesses that he has just killed his mother cannot complain if the doctor calls the police. Most cases are not as obvious as this. A policeman appears in the occupational health department, asking for information about employees in connection with a burglary during which one of the thieves escaped by jumping off a wall into some bushes. The doctor knows that one particular worker has recently been treated for scratches and cuts which he said were the result of a weekend's gardening. The doctor is under no legal obligation to tell the police anything. Is it in the public interest to be frank, when the employee may be completely innocent? The answer is probably that confidence should be preserved, since there is no imminent danger to others.

The concept of public interest has been widened by the courts in recent years. Mr Evans, a scientist formerly employed by Lion Laboratories, was permitted by the Court of Appeal, against the wishes of his former employer, to publish details in the Press of secret technical aspects of the Lion Intoximeter, because the reliability of the device was a matter in which the public were justifiably concerned (*Lion Laboratories* v. *Evans* (1984)). On the other hand, the courts in 1987 granted an injunction against *News of the World* preventing it from publishing hospital records identifying two NHS doctors who were suffering from AIDS (*X* v. *Y*). 'The public in general and patients in particular are entitled to expect hospital records to be confidential and it is not for any individual to take it upon himself to breach that confidence.' In the latter case there was little actual danger to patients. It is highly unlikely that a health professional will transfer the AIDS virus to his patient in the course of practice. After this case, the GMC and the UKCC issued guidance to members of the professions who are carriers of the virus. They should seek medical advice and act on it. Only if they refuse to take the precautions recommended to them should their medical advisers report the matter to third parties (see Chapter 2).

A recent case in which the doctor's duty of confidence was considered by the courts was *W* v. *Egdell* (1990). An independent psychiatrist had been asked by solicitors acting for W, a patient detained in a secure hospital, to examine him with a view to using his report to support an application to a mental health review tribunal for W's discharge or transfer to a regional secure unit. W, diagnosed as a paranoid schizophrenic, had killed five people ten years before. Dr Egdell wished the tribunal to know that he opposed W's application, because he believed him to be a psychopath, but found that his adverse report on W's condition had been suppressed by the solicitors, though they subsequently withdrew the tribunal application. The psychiatrist felt that it was his public duty to communicate his views about W to the authorities, so that they would be kept with his records and be available on any subsequent application. He sent his report, against the wishes of the patient, to the medical director of the hospital and the Home Office. The Court of Appeal dismissed W's action for breach of confidence. Dr Egdell's duty to the public justified his actions.

Occasionally, a doctor or nurse may be confronted with an employee who knows that his job is endangering his health (for example because he has developed an allergy or is putting too much strain on a damaged joint), but refuses to allow the health professional to alert third parties. In the case of statutory medical examinations, the Appointed Doctor or Employment Medical Adviser will be obliged to tell the employer and the HSE that the employee is unfit. Otherwise, if there is no risk to third parties, both law and ethics probably dictate silence. The worker should be fully informed of the risk that he runs. For the protection of the doctor or nurse, he may be asked to sign a statement that it is at the worker's request that his condition is being kept from management. The Faculty of Occupational Medicine in its *Guidance on Ethics* (1993) states that occupational physicians should appreciate that when advising on the nature or extent of a work-related health risk, they should not presume to decide for others whether or not that risk is acceptable.

One of the most difficult dilemmas for the doctor or nurse is to be confronted with a state of affairs where he has to choose between competing interests. Suppose that a doctor in the course of a confidential consultation with an employee discovers that he is an alcoholic and is likely to be operating machinery while inebriated. If he keeps confidence an accident may happen in which a third party may be injured; if he tells the personnel department that the man is unfit, the employee may be sacked and the rest of the workforce become much more wary of confiding in the doctor in the future.

No-one can give perfect advice on how to deal with such a dilemma, but the following pointers may prove valuable:

(1) The patient may be persuaded of the dangers of the situation and may give permission for the truth to be revealed, or at least be willing to ask for a transfer.
(2) If the employer has enlightened policies, it may be possible for the worker to keep his job in the long term if he successfully undergoes treatment. The occupational health department can prove influential in persuading management to adopt such policies.
(3) Influential professional bodies recognise that in these cases it may be permissible to breach confidence. The Faculty of Occupational Medicine's *Guidance on Ethics* says this:

> 'Occasionally, an occupational physician ... may find that an individual is unfit for a job where the safety of other workers or of the public is concerned. He should then take great care to explain fully why he thinks the disclosure of unfitness is necessary. If sufficient time is taken in explanation, there is rarely difficulty in obtaining agreement. When this is not obtained the occupational physician is faced with an ethical dilemma. No firm guidance is possible, and each situation must be considered on its merits. On such occasions it may be useful for the occupational physician to discuss the issue with an experienced colleague. Ultimately, the safety of other workers and the general public must prevail as one of the exceptions to the duty of confidentiality.'

3.8 Legal obligation to disclose information

A number of different statutory provisions compel disclosure of information to a public body. The principal examples relevant especially to occupational health practice are as follows:

(1) The reporting of notifiable infectious diseases (Public Health (Control of Disease) Act 1984).
The following diseases must be reported by the doctor to the local authority: acute encephalitis, acute meningitis, acute poliomyelitis, anthrax, cholera, diphtheria, dysentery, infective jaundice, Lassa fever, leprosy, leptospirosis, malaria, Marburgh disease, measles, ophthalmia neonatorum, paratyphoid fever, plague, rabies, relapsing fever, scarlet fever, smallpox, tetanus, tuberculosis, typhus, viral haemorrhagic fever, whooping cough, yellow fever and food poisoning.
  AIDS is not a notifiable disease, although certain provisions of

the Public Health Act have been extended to AIDS, giving power to a magistrate on the application of a local authority to order the compulsory detention of an AIDS sufferer in hospital if he would not take proper precautions to prevent the spread of the virus if allowed to go free. There are special regulations obliging confidence to be preserved in cases of sexually transmitted diseases. The NHS (Venereal Diseases) Regulations 1974 require health authorities to take all necessary steps to secure that any information capable of identifying an individual obtained with respect to persons examined or treated for any sexually transmitted disease shall not be disclosed *except for the purpose of communicating the information to a doctor or someone working under his direction in connection with the treatment of persons suffering from the disease or the prevention of its spread*. It will be seen that, contrary to the practice of hospital sexually transmitted disease clinics, the regulations allow the patient's GP to be notified. It is the general law of confidence which forbids disclosure to the GP without the patient's consent, except in a situation of overriding public interest.

(2) Notification of drug addicts (Misuse of Drugs (Notification of and Supply of Addicts) Regulations 1973).
Doctors are required to send to the Chief Medical Officer at the Home Office particulars of persons whom they consider or suspect to be addicted to certain controlled drugs, i.e. cocaine, dextromoraride, diamorphine, dipipanone, hydrocodone, hydromorphone, levorphanol, methadone, morphine, opium, oxycodone, pethidine, phenazocine and piritramide.

(3) Notification of acts of terrorism (Prevention of Terrorism (Temporary Provisions) Act 1989).
It is an offence for any person without reasonable excuse to fail to disclose to the police information which would be of material assistance in preventing terrorism or in securing the arrest of any person involved in an act of terrorism.

(4) Notification of accidents and dangerous occurrences at work and of work-related disease (Reporting of Injuries, Diseases and Dangerous Occurrences Regulations (RIDDOR) (1985)).
There have been laws about the notification of accidents in the workplace for many years but it was only in 1980 that one set of regulations applying to all premises where people were employed was introduced. These required notification to the local authority or Health and Safety Executive (HSE) not only of major accidents but also of dangerous occurrences, i.e. incidents which could have

led to serious injury but by good fortune ended up as 'near-misses'. The introduction in 1983 of the statutory sick pay scheme, which shifted the burden of short-term sick pay from the DSS to the employer, meant that there was no longer an automatic procedure for notifying the Government of minor injuries at work. Also, it was decided that it was important to try to identify the occupational risks of contracting diseases through the collection of accurate statistics. New regulations made in 1985 which place the duty to report on the employer or other person in control of work premises require notification of the following:

(i)   Major accidents. These are accidents causing death or (a) fracture of the skull, spine or pelvis; (b) fracture of any arm, wrist, leg or ankle bone (other than one in the hand or foot); (c) amputation of a hand, foot, finger, toe or thumb; (d) loss of sight or other eye injury; (e) electrical shock injury or burns or loss of consciousness; (f) loss of consciousness from lack of oxygen; (g) decompression sickness; (h) acute illness or loss of consciousness from the absorption of any substance; (i) acute illness from exposure to a pathogen or infected material; (j) any other injury which results in the person being admitted immediately into hospital for more than 24 hours. If an employee's death occurs due to a reportable accident within a year of that accident it must be reported.

(ii)   Dangerous occurrences. These are listed in Schedule 1 to the regulations and include the collapse of, the overturning of, or the failure of any load-bearing part of any lift, hoist, crane, derrick or mobile powered access platform, the explosion, collapse or bursting of any closed vessel which might have caused death or serious injury, an explosion or fire resulting in a stoppage for more than 24 hours, the uncontrolled or accidental release or the escape of any substance or pathogen in circumstances likely to cause death or serious injury and any ignition or explosion of explosives where the ignition or explosion was not intentional.

(iii) Minor accidents causing a person to be incapacitated for work for more than three consecutive days (but not falling within the definition of a major accident).

(iv)   Industrially linked diseases. Schedule 2 to the regulations lists 28 forms of illness in Column 1 with, opposite each of these, a list of occupations or work-processes in Column 2. If an employee contracts a Column 1 disease while working in a Column 2 process it is notifiable. Some examples are mesothelioma, lung cancer or asbestosis contracted while working with asbestos, hepatitis contracted while working with human blood products or body secretions or excretions, bone cancer or blood dyscrasia while working with ionising radiation and vibration white finger while using various hand-held tools or percussive drills or holding material being worked upon by pounding machines in shoe manufacture.

A case of disease in an employee must be reported only if a written diagnosis has been received from a doctor, for example, on a medical certificate. Doctors are asked to use the common descriptions of each disease set out in the regulations so that the employer can easily recognise when a report must be made. If a person receiving training who is not an employee is involved, the training body must report, and the self-employed should report in respect of themselves. Reporting a case of disease does not necessarily signify that it was caused by work. The employer does not have to report if the employee does not work in a listed job even if he knows that he worked in such a job in the past.

Note that major accidents and dangerous occurrences must be notified to the local authority environmental health department (responsible for offices, shops, warehouses and residential accommodation) or the Health and Safety Executive by telephone, followed by a written report within seven days; minor accidents and industrially-linked diseases are to be notified on the standard form. Records must be kept and the authorities may ask for further information of notified matters (which may include medical records). If the employer can prove that he had taken reasonable steps to have all notifiable events brought to his notice but that nevertheless he was not aware of the notifiable accident, event or disease he will not be guilty of any offence; for this reason it is important to institute a reliable and comprehensive reporting system among employees. There is evidence of considerable non-compliance by employers, and the regulations are under review. Since 1990 a special supplement to the Employment Department's Labour Force Survey has disclosed just how inaccurate are the RIDDOR statistics. The Health and Safety Commission has estimated that no more than 30 per cent of work-related injuries are being reported. The danger of failure to report is, of course that it produces inaccurate statistics which may conceal serious dangers or suggest inappropriate trends.

Some employers ask occupational health departments to take responsibility for reporting accidents and disease. This should be resisted, since it may bring the duty of confidentiality of the health professional towards the workers into conflict with his duty to the employer.

3.9 Legal obligation to reply to questions

Sometimes, there is an obligation to give information, but only if an official body requests it. The main category of cases falling under this head involves the giving of evidence in the course of legal proceedings, or where legal proceedings are pending. English law does not, unlike some other systems, confer any special privilege on health professionals to refuse to disclose information confided in them by their patients *before* legal proceedings are in prospect. The only statements privileged from disclosure even in the highest courts are those which attract *'legal professional privilege'*, that is either communications between a client and his professional legal adviser or communications between a third party and the client or his lawyer, *where they were made in contemplation of litigation and the dominant purpose was to prepare for litigation*. Thus, a doctor who is asked by solicitors to give expert medical evidence in connection with a claim for compensation and prepares a report for that purpose may find that it is never produced because it is not as favourable as was wished. The privilege here is not that of the doctor, but of the lawyer's client. If an employer instigates an inquiry into an accident at work, partly because he anticipates litigation, but partly because he wishes to make a full investigation to ensure that the incident is not repeated, the subsequent report is not privileged against disclosure, because litigation is not the sole or dominant purpose for which the document was prepared (*Waugh* v. *British Rail* (1980)).

There is no general legal obligation on anyone to answer questions posed by the police, or give them access to medical records, even where they are investigating a serious crime, though if a deliberate lie is told or money is taken for suppressing information a criminal offence may be committed. One exception is the Road Traffic Act which obliges answers to police questions where they are investigating a serious driving offence. Under the Police and Criminal Evidence Act 1984, a search warrant to give the police access to medical or counselling records or samples of human tissue or fluid taken for the purposes of medical treatment may be obtained from a circuit judge but only in very exceptional cases. In *R* v. *Cardiff Crown Court, ex parte Kellam* (1993), it was held that the court should not order disclosure of records relating to the times of admission, discharge and leave of patients in a mental hospital, which the police wanted to check the movements of a mental patient suspected of murder. Should a doctor demand a search warrant before giving information to the police? It is probably advisable where there is no element of urgency, but in a case of danger that further crimes may be committed, the courts, the GMC and the UKCC all sanction breach of confidence. If a prosecution actually proceeds to trial, a doctor or nurse could be called as a witness and ordered by the judge or the magistrates to answer questions on oath. Professional ethics would not in law justify a refusal to answer, which would constitute a contempt of court, as would a refusal to comply with a search warrant.

The police occasionally assert that they have a right to search a doctor's manual or computer records without a warrant where they have entered premises in order to carry out a lawful arrest or with the permission of the occupier. This is probably true where they have reasonable grounds for believing that they will find evidence of a crime which must be seized in order to prevent it being concealed, lost, altered or destroyed.

Crimes against the health and safety of the workforce are policed by the HSE and the local authorities. These public officials have special powers of enforcement which include the power to demand information, and exceed the powers of the police. Section 20 of the Health and Safety at Work Act 1974 (HSWA), discussed in Chapter 5, gives them power in the course of an examination or investigation to inspect and take copies of books and documents without a warrant and without the permission of the occupier, as long as they do so for the purpose of enforcing safety legislation. There is no exemption for medical records and all documentary records, whether or not required by statute, are included. The inspector may require any person whom he has reasonable cause to believe to be able to give any relevant information to answer questions and to sign a declaration of the truth of his answers. Such a statement will not be admissible against its maker in any legal proceedings. The privilege against self-incrimination means that in any event the person actually suspected of a crime has the right to remain silent, but s. 20 allows the inspector to compel a *witness* to give evidence when he might otherwise decline for fear of reprisal from his employer. An occupational health professional might be ordered to make a statement about the medical condition of a worker. If he declined, he could be prosecuted for a criminal offence under s. 33 HSWA.

In addition, the Health and Safety Commission (HSC) has power, with the consent of the Secretary of State (which may be given in general terms), to serve a notice demanding that information be furnished either to the HSC or the HSE of matters needed to carry out their statutory functions. Although this information is not to be disseminated to the public as a whole, it may be given to local authorities and to the police. It may also be disclosed in legal proceedings. The inspectors *shall*, in circumstances where it is necessary to do so for the purpose of keeping persons (or their representatives) employed at any premises adequately informed about matters affecting their health, safety and welfare, give to them or their representatives factual information relating to the premises or anything done therein and information about any action they propose to take.

The Environment and Safety Information Act 1988 imposes a duty on the authorities empowered to issue improvement and prohibition notices under the Health and Safety at Work Act to maintain a register of such notices open to public inspection. An entry may be modified to protect trade secrets or secret manufacturing processes.

In 1992 the Environmental Information Regulations came into force. The HSC/HSE will make environmental information available on request in those areas where they have environmental responsibilities. These areas are pesticides, new substances, genetically modified organisms, control of industrial major accident hazards, onshore and cross-country pipelines and polychlorinated biphenyls and polychlorinated terphenyls. The regulations exempt information affecting national or public security, commercial confidentiality, personal privacy and information voluntarily supplied.

A Coroner's Court also has power to order disclosure of medical records or other confidential medical information.

If the employee is dismissed because the employer determines that he is medically unfit, it is possible that the OH physician may be called as a witness and ordered to produce records before an industrial tribunal. The same may occur when an employee is suing his employer for negligence or breach of statutory duty in the High Court or County Court, or claiming social security benefits before a Social Security Appeal Tribunal or Medical Appeal Tribunal. All these have the power to order disclosure and the physician or nurse who disobeys could be punished for contempt of court. However, the judge or tribunal chairman has a discretion about the admission of evidence. In one case the House of Lords refused to order the NSPCC to disclose the name of an informant, lest others be deterred from reporting suspected cruelty (*D* v. *NSPCC* (1978)). The Faculty of Occupational Medicine advises that in civil cases, unless the patient has consented to disclosure, the physician should wait for a formal court order, 'but physicians should not be misled by solicitors who state that an application for a court order has been made'.

3.10   Pre-trial disclosure in civil proceedings

There is a distinction between pre-action disclosure (before the writ is issued) and pre-trial disclosure, known as discovery (after the action has commenced but before it has come to court). Pre-action disclosure of medical records is often made as a result of an application under section 33 of the Supreme Court Act 1981 in an action for damages for death or personal injury. If a plaintiff is able to show that it is likely that he may have a case for damages for some injury, he may ask the court for an order against a possible defendant (even before the issue of a writ). This procedure is not available in industrial tribunals, nor is the procedure under s. 34 of the same Act whereby disclosure may be ordered of records in the possession of a third party, like a doctor or a hospital (after the writ has been issued). The court now has a discretion whether to confine the order to disclosure to a named physician or lawyer nominated by the plaintiff, rather than the patient in person. This power may be exercised if the medical details are particularly distressing (as where the patient is dying), or very technical.

Where a complaint of unfair dismissal has been made to an industrial tribunal, either party can request an order that the other produce relevant documents for perusal before the hearing. Since the OH professional would not be a party to the unfair dismissal complaint, the only way he could be compelled by the employer to give evidence or produce his medical records without the consent of the employee would be to seek a witness order requiring him to produce them at the hearing. The problem would then arise that the employer would be unable to show that he had relied upon the medical evidence in making his decision to dismiss. Employers are limited to the evidence known to them at the date of dismissal when they seek to prove that they acted fairly (*Polkey* v. *Dayton Services* (1988) (see further Chapter 8)).

Where the Access to Health Records Act 1990 applies, the patient or his representative, and, where the patient has died, any person who may have a claim arising out of the patient's death, has a right to see health records.

3.11 *Confidential information in the courts*

Courts are not eager to order disclosure of confidential information unless it is necessary in the interests of justice. In a case in 1979, a woman civil servant complained that she had not been promoted because of her sex and her trade union activities; she asked the industrial tribunal to order her employer to produce not only her own confidential records but also those of fellow employees who had been promoted, for comparison. The House of Lords held that the tribunals should, if necessary, censor confidential files before allowing the other party to see them, so that only relevant information is revealed (*Science Research Council* v. *Nassé*). Needless to say, the doctor or nurse should not assume the role of censor, though courts will be sympathetic to requests to hold back records if good reasons are given.

A case which involved occupational health records raised an interesting debate. Mr Nawaz was an employee of the Ford Motor Co. After 18 months' absence from work during which he had provided his employer with certificates from his doctor saying that he was unfit, he was asked to submit to an examination by the employer's OH physician. The latter decided that Nawaz was fit, but the GP continued to certify him as unfit. Nawaz agreed to see a consultant nominated by the employers. The OH physician gave the consultant information about Nawaz and his job; the consultant eventually reported that in his opinion Nawaz should recommence work because he was fit to do so. When the employee continued to refuse, he was dismissed for unauthorised absence. On an application to an industrial tribunal for unfair dismissal, Nawaz asked for disclosure of all his medical records, including the correspondence between Ford's doctor and the consultant and the consultant's report in full. The company objected that the managers who had made the decision to dismiss had never seen those records – they had respected medical confidence and acted on the experts' conclusion that Nawaz was fit. Since the employer's duty under the Employment Protection Act is only to act as a reasonable employer on the available evidence, the issue before the tribunal was not whether Mr Nawaz was in fact fit for work, but whether it was reasonable for management to act on that assumption, so that detailed medical information was not relevant and should not be produced.

The Employment Appeal Tribunal disagreed. Discovery must be granted if it is necessary to dispose fairly of the action. Even though management are not expected to be able to assess the merits of medical argument, they are not absolved from carrying out, through their medical agent, the proper investigation which is required in any case of dismissal. Nawaz needed to examine the medical correspondence to be sure that the consultant had been properly informed and that his conclusions justified the advice that Nawaz was fit. In other words, it was possible that Ford's doctor through prejudice had written that Nawaz was a well-known malingerer and should not be trusted; only if all the records were produced would it be clear that he had acted fairly (*Ford Motor Co. v. Nawaz* (1987)).

The power to order discovery is only conferred by law on courts, tribunals and other similar bodies, like the Health Service Commissioner (Ombudsman). The Advisory, Conciliation and Arbitration Service (ACAS) does not have such powers, nor do purely domestic inquiries set up by an employer to investigate any matter. If there is doubt about whether a doctor or nurse is obliged by law to produce confidential records, advice should be sought from the professional organisations.

3.12 *Giving evidence in court*

If the case comes to trial, and most civil cases are settled out of court, often at the eleventh hour, the doctor or nurse may be summoned to appear as a witness. They may be there in one of two capacities. Most commonly, an OH physician or nurse will be called because he has been involved in the treatment of a patient whose health is an issue before the court. In these circumstances, the evidence will be drawn from personal knowledge of the course of the patient's illness. In some cases, however, the health professional may be called as an expert witness, not because he has had any clinical contact with the patient prior to the legal proceedings, but because he can advise the court about technical matters not within the judge's competence. In the former case, if the judge insists, he may be ordered to answer questions and may be punished for contempt if no answer is forthcoming. (When the health professional feels that information about a patient is particularly sensitive, he should ask the judge to exercise his discretion by allowing the information to remain secret; if necessary, the judge may first peruse the evidence in private. The judge will always respect the desire to preserve confidence if the interests of justice permit.) In the latter case, the expert will only have become involved after legal proceedings were mooted, so that he will only be summoned to give evidence if he has volunteered to do so and the patient has agreed, and no breach of confidence will arise. The other side will be unable to obtain an order compelling disclosure of an expert medical report which has been rejected by the party who requested it as unfavourable to his case. Such a document is covered by legal professional privilege as a communication made in contemplation of litigation (see above).

An example was the case of a plaintiff, Mr Brooks, who worked in a cotton spinning mill for twenty years and was eventually diagnosed as suffering from byssinosis. He claimed damages from his employers, J. and P. Coates Ltd, because, he alleged, they had been negligent in not providing sufficient ventilation. The employers' main defence was that Mr Brooks's symptoms were caused by smoking not by cotton dust and they produced the evidence of confidential medical records held by his GP which showed only a diagnosis of bronchitis caused by cigarette smoking. Three consultant physicians brought into the case only after Brooks decided to take legal proceedings and who examined him at his request gave expert evidence for the plaintiff that his disability was to some extent caused by cotton dust. This diagnosis rested crucially on the plaintiff's account of suffering increased breathlessness on Mondays after a break from work (a classic indication of byssinosis). There was no mention of this symptom in the GP's records. The judge accepted the plaintiff's account and gave him damages to compensate for 50 per cent of his disability (the other 50 per cent was held due to cigarettes) (*Brooks* v. *J. and P. Coates Ltd* (1984)). Note that the GP was compelled to produce his records, but that the court could not have compelled the consultants' reports to be produced if the plaintiff had wished to suppress them because they were covered by legal professional privilege.

When giving evidence, the witness should be careful to avoid speculation and should stick to the facts. If he does not understand a question he should ask for it to be repeated. If he does not know the answer to a question he should say so. Remember that it will often be the job of the lawyer for the other side to cast doubt on the medical evidence and that this can be done by making a witness angry and flustered.

Because, as has been shown, any records, however private, may end up being read out in open court, it is important not to indulge in jokey comments and to make records as clear as possible. It is also an important protection for the patient who may still be working for the employer after the health professional has moved on to another post.

3.13   The patient's right to know

Where it is the patient who is demanding his own records, confidentiality is irrelevant, because it is the *patient's* confidence which the law protects. The right of subject access should be given, and not merely to encourage the paranoid. A case reported in the BMA's *News Review* concerned Helen Mann, a young woman who moved home and transferred to a new GP. When she visited her new doctor for the first time he began to discuss her addiction to drugs which appeared on her records. As she had never taken any illegal drug, she was totally nonplussed. Investigation revealed that a heroin addict of about the same age had been using her name. It was only by proving that she was abroad at the time the counterfeit Helen had been seen that she persuaded the doctors to amend the notes. No real harm had been done, but if the patient had not changed doctors she might have been refused jobs, mortgages or insurance because of a bad medical report without ever knowing the source of her lack of success.

The right to receive information is as valuable as the right to see medical reports and records. In the following situations the patient has a legal right to be told details and/or to see his own medical records:

(1) Where the patient is being asked to give consent to an examination or procedure, he must be given sufficient information to enable him to give an informed consent. The doctor must satisfy the *Sidaway* test (see Chapter 2), i.e. he must tell the patient what a reasonable physician would have told him in all the circumstances of the case.

(2) Where the patient must be given information to warn him or her of the need for treatment. For example, an occupational physician institutes cytological smear testing of female employees. If he finds any adverse signs, he will be liable in negligence if he does not take reasonable care to notify the employee concerned of the need for further investigation.

(3) Where personal information relating to a living individual is recorded in a form in which it can be processed by equipment

operating automatically in response to instructions given for that purpose, e.g. in a computer, the Data Protection Act 1984 gives the right to the individual who is the subject of the data (the data subject) to see it on request after a reasonable time and on payment of a reasonable fee (see below). For our purposes, the most important exceptions are those in the regulations relating to health and social work records, but there are also exceptions for data held for the prevention or detection of crime, the apprehension or prosecution of offenders and the assessment or collection of any tax or duty.

(4) The Access to Medical Reports Act 1988 gives patients the right to see copies of medical reports requested from their doctors for employment or insurance purposes (on payment of a reasonable fee). The Act refers to 'a report relating to the physical or mental health of the individual prepared by a medical practitioner *who is or has been responsible for the clinical care of the individual'*. 'Care' includes examination, investigation and diagnosis for the purposes of, or in connection with, any form of medical treatment. On its face, this excludes reports of independent medical examinations and pre-employment screening by the occupational physician (a pre-employment report by the GP or consultant treating the job applicant would fall squarely within the provisions of the Act). More doubtful will be reports by OH physicians on existing workers, since in some cases but not others there will have been a clinical care relationship with an individual worker. Surely it would be invidious to single out a minority of the workers: in my view, it would seem good practice to comply with the Act in all cases.

This legislation will give patients a chance to correct errors (or, if the doctor is unwilling to amend the report, to attach to it a statement of the patient's views), and to withdraw consent to reports being sent if they object to them. A doctor who has been notified that a patient desires access to his report before it is sent to the employer or insurance company will have to wait 21 days. If the patient has not by then taken steps to arrange access, the doctor can supply the report without the patient seeing it. Copies of medical reports supplied for employment or insurance purposes must be retained for at least six months and must be available on request to the person to whom the report relates if application is made within six months of the report having been supplied.

The doctor will be able to withhold a report from his patient if it would in his opinion be likely to cause serious harm to his physical or mental health or the physical or mental health of others, would indicate the intentions of the practitioner in respect of the individual, or would reveal the identity of or information about another person without that person's consent (except a health professional involved in the care of the individual). In this event, the patient will have to be notified that the report is to be kept from him, so that he can decide whether he still wishes to consent to it being supplied to an insurance company or current or prospective employer. The Act provides that a complaint of a breach of the Act may be made to the County Court, which may order the employer, insurance company or doctor to comply.

(5) The Access to Health Records Act 1990 gives patients access to non-computerised health records made after 1 November 1991 and to information recorded earlier when it is necessary in order for the patient to understand what was written later. Both NHS and private health records are covered. Patient access outside the scope of the Act is at the health professional's discretion. The High Court confirmed in 1993 that a patient does not have a right of access to records made before the Act came into force (*R* v. *Mid-Glamorgan FHSA, ex parte Martin*). Health record means a record which relates to the physical or mental health of an individual and has been made by or on behalf of a health professional *in connection with the care of that individual*. This should be contrasted with the Access to Medical Reports Act which refers to responsibility for clinical care and probably reflects the fact that the Act applies to health records kept by all kinds of health professionals. Nurses, dentists, pharmacists, chiropodists and art therapists are all included. Care is defined as including examination, investigation, diagnosis and treatment. Because of the difference in wording, it is possible to argue that the Act applies to all occupational health records, including records of pre-employment examinations (but not questionnaires). Since one purpose of the Act is stated to be the correction of inaccurate health records, it is my view that the inclination of the courts would be to give a broad interpretation.

There are the usual exclusions. Access shall not be given to any part of a health record which, in the opinion of the holder of the record, would disclose information likely to cause serious harm to the physical or mental health of the patient or of any other individual, or information relating to or provided by an individual, other than the patient, who could be identified from that information.

Applications for access must be made in writing to the holder:

the patient's GP, the health authority in the case of hospital records, or the head of the occupational health department. Complaints about non-disclosure may be made through NHS complaints procedures or to the courts. Though the Act is supposed to run parallel with the Data Protection Act, there is no public official equivalent to the Data Protection Registrar with power to police the 1990 Act. Applications may be made by the patient, someone authorised in writing to make the application on his behalf, the parents of a child, the person appointed by the court to manage the affairs of an incompetent and, where the patient has died, his personal representative and any person who may have a claim arising out of his death.

An important section for the OH professional is s. 9 which renders void any term or condition of a contract which purports to require an individual to supply any other person with a copy of his health record. Without this, employers might have imposed a contractual obligation on workers to obtain and disclose health records at the employer's request. However, the section will not prevent employers from asking job applicants to provide health records at the pre-employment stage (see Chapter 4).

(6) The Access to Personal Files Act 1987 gives the right of subject access to local authority housing and social services records. Information held by social services which originates from health professionals should not be revealed without first consulting the relevant health professional. Damage to the subject's health and revelations about a third party who does not consent to disclosure justify withholding records or part of them.

(7) Statutory regulations sometimes direct that individuals must be given information about themselves. Examples in the field of health and safety at work are those relating to the control of lead and ionising radiations, and also the COSHH Regulations. The pattern of these regulations is to impose a legal requirement on both employer and employee to undertake regular testing by doctors appointed by the Employment Medical Advisory Service (EMAS). The detailed record of these tests identified by the name of the individual worker is a confidential document which should not be disclosed to the employer or the general practitioner without the consent of the individual, although it must be made available for inspection by EMAS. The law requires that the Appointed Doctor inform the individual of his fitness to work with the dangerous substance, and notify the employer that he has undertaken the necessary examinations. The outcome of the examination may only

be made known to the employer if the employee consents, but the employer and EMAS *must* be notified of persons certified as unfit, or fit only under specified conditions. Group results without individual identification can be given to the employer and to other interested parties when relevant to the provision and/or effectiveness of preventive measures.

The Faculty of Occupational Medicine recommends that the individual has the right to be informed of the findings of any clinical examination, whether or not required by statute, but that these should not be disclosed to a third party (including his own doctor) without the individual's informed consent.

It may be important for work representatives to be told about possible hazards. Though they should not given access to the medical record of an individual without his consent, s. 28 of the Health and Safety at Work Act allows the HSE to reveal information to employees or their representatives which is necessary to keep them adequately informed about matters affecting their health, safety and welfare. There is nothing in s. 28 which excludes the disclosure of confidential medical information, though in practice it would be most unlikely that identifiable patient information would be revealed by this method. Section 28 also permits an inspector to furnish any person who appears to him likely to be a party to civil proceedings with a written statement of relevant facts observed by him in the course of the exercise of his statutory powers. The Safety Representatives and Safety Committees Regulations 1977 provide that safety representatives shall be entitled to inspect any document which the employer is required by safety legislation to keep, except a document consisting of or relating to any health record of an identifiable individual.

Doctors believe that if Doctor A writes to Doctor B about a patient, Doctor B must refuse to show the letter to the patient in question because it would be a breach of confidence. The law is that in circumstances where the patient has a right to know, such a letter must be revealed. As has been previously stated, it is the *patient's* confidence which is protected.

3.14   Data Protection Act 1984

Medical records consisting of the results of regular tests are increasingly computerised. The law of confidence as it relates to computer data is now codified in the 1984 Act. The absurdity of having different rules for different kinds of records has been stressed by many commentators. The Act imposes increased legal liabilities on those who control personal data recorded in a form in which it can be processed by equipment operating automatically in response to instructions given for that purpose. Word processors are included unless they are used only for preparing the text of documents. 'Personal data' means data consisting of information which relates to a living individual who can be identified from that information (or from that and other information in the possession of the data user). You cannot avoid the Act by keeping the key to information stored on the computer in manual form. Expressions of opinion about an individual are included in the term data, so that records of biological testing and more traditional doctor/patient consultations will both be covered (in practice the first are more likely to be computerised than the second).

The Act requires the 'data user' to comply with a number of statutory requirements. The data user is defined in the Act as a person who holds data, that is the person (either alone or jointly or in common with other persons) who controls the contents and use of the data. Although in the normal case the employing organisation is the data user rather than individual employees, the special position of occupational health staff means that it is they who are in control and on whom the statutory obligations fall.

These are as follows:

(1) The duty to register with the Data Protection Registrar.
A failure to register the holding of personal data is a criminal offence. The data user must register his name and address, a description of the data held by him and the purposes for which the data are held, a description of the sources from which the data are to be obtained, a description of those to whom he may wish to disclose the data and any foreign countries to which he may wish to transfer the data. Registration, for which payment is required, will have to be periodically renewed. Anyone who knowingly or recklessly holds personal data other than as entered in the register or discloses it to any person who is not described in the register, except when that disclosure is exempt (see below), is guilty of a criminal offence. 'Work-related' records held on home computers for convenience must be registered; only private information like personal tax returns falls outside the Act.

(2) The duty to observe the data protection principles, which include the duty to obtain and process information fairly and lawfully, the duty to specify the lawful purpose for which data are held and not to use or disclose it in a manner incompatible with that purpose, the duty to keep only accurate, relevant and up-to-date information, the duty to discard data when it is no longer necessary to keep it, the duty in some circumstances to give the subject of the data access to it, and the duty to take appropriate security measures against unauthorised access to, or alteration, disclosure or destruction of personal data and against accidental loss or destruction of personal data. The Faculty of Occupational Medicine recommends that as a precautionary measure a regular updated copy of the computer records should be taken and kept in a separate and secure location. No unauthorised software should be used on the system to prevent the risk of computer viruses being introduced which can corrupt information. A policy statement detailing procedures for security and confidentiality should be prepared, and all staff with access to the occupational health unit should receive adequate training and sign the policy statement to acknowledge their responsibilities.

(3) The duty to allow the data subject access to data about himself. The right of subject access is defined as the right to be informed whether data are being held and to receive a copy of the data in a form in which it can be understood, within a reasonable time (usually 40 days). The Act lays it down as a general principle that the subject of data should be given a copy of those data on request, which must be made in writing. The data user is entitled to charge a small fee. He should check that the applicant is in fact the data subject, or someone acting with his authority. Also, there are number of exceptions. The researcher who holds personal data on his computer for research purposes (unless those data cannot be identified by any means) must register under the Act, but is exempt from the subject access provisions if the research is not made available in a form which identifies the data subjects. Communications between client and lawyer are exempt if covered by legal professional privilege.

There are special rules about medical and social work records in statutory regulations which came into force in November 1987. The order relating to medical records (Data Protection (Subject Access Modification) (Health) Order (1987)) applies to personal data consisting of information as to the physical or mental health of the data subject if the data are held by a health professional or the information in the data was first recorded by or on behalf of a health professional (personnel and insurance records including medical information are included). Subject access may be refused if either it would in the health professional's opinion be likely to cause serious harm to the physical or mental health of the data subject or it would reveal the identity of another individual, who does not wish to be identified, as the subject or the source of information. An entry in a patient's record that his wife had reported him to engage in bizarre behaviour might be withheld either because it would harm the patient to know what had been said or because the wife did not want her husband to realise what she had told the doctor.

Suppose that the source of the 'sensitive' information is another health professional? A GP writes to an OH physician in confidence that his patient, about whom the latter has enquired, is HIV positive. This is entered by the OH physician in his computer records. The patient demands access to his occupational health record. Compliance will expose the GP's revelations and he is therefore opposed to it. The patient is already aware of his condition so his health will not be damaged by disclosure. The regulations quite clearly state that the data must be revealed. The data user is not excused from supplying data which reveal information which relates to a health professional who has been involved in the care of the data subject, or was supplied to the data user by him in his capacity as health professional. Doctors who complain that this rule runs counter to their professional etiquette should remember that the patient is asking for his own information and that it is *his* confidence which the rules of law and ethics seek to protect, not that of another doctor.

Who is empowered to decide whether medical records should be released to the patient? The obligation to disclose is on the data user. Assuming that occupational health records are under the control of OH doctors and nurses, they will have to make the decision within the regulations. The same applies to general practitioner records since the GP will be registered as the data user of his patient's computer records. Hospital records are under the control of the health authorities who are registered under the Act. The regulations provide that, in the case of a data user who is not a health professional, data shall not be revealed without consulting the health professional (defined as including doctors, dentists, opticians, pharmacists, nurses, midwives, health visitors, and other paramedical staff) who is currently or was most recently responsible for the clinical care of the patient, or if not available a suitable professional with the qualifications and experience to advise. The duty is to consult, not necessarily to obey.

The regulations envisage that medical records may have to be censored by removing the sensitive information and disclosing the rest. From the patient's point of view, it will be virtually impossible to challenge a decision to withhold data, because he will not be aware that anything is being kept from him.

The Act provides a number of sanctions against violation of the data protection principles. The Registrar may take steps to investigate alleged violations which could lead to a withdrawal of registration, subject to an appeal to the Data Protection Tribunal. An individual who is the subject of personal data who suffers damage by reason of its inaccuracy may have a right to compensation for damage and distress, but it is a defence for the data user to prove that he took reasonable care to ensure the accuracy of the data at the material time (s. 22). Suppose that an occupational physician records that he has been told by the shop steward that one of the drivers is epileptic. If the entry accurately records that the information has been received from a third party and, once the driver notifies the physician that he regards the information as incorrect or misleading, an indication to that effect is included in the data, no action for damages will lie. Of course, if the driver has not been given access to his records he will remain ignorant of the damaging allegations.

An individual who is the subject of personal data and who suffers damage by reason of the loss of the data, the destruction of the data without the authority of the data user, or the disclosure of the data or access to it without the authority of the data user, shall be entitled to compensation for damage and distress (s. 23). Note that this section deals with failure to keep data secure: the loss or disclosure of the data without the permission of the person holding it, not with disclosure to a third party without the permission of the subject of the data. If, for example, there is a secretary in the occupational health department who is a well-known gossip and is given no instructions about the need to keep medical details confidential, an employee who finds that no-one will work with him because they have discovered from disclosure of his computer record by the secretary that he is an HIV carrier would have an action for damages under this provision against the doctor or nurse, the data user, even though the secretary is not employed by them but by the employing organisation. But the section would not apply, though there would clearly be an action at common law, to disclosure by the secretary on the instructions of the data user of the same information to the personnel department. In proceedings brought against any person under s. 23 it is a defence to prove that he had taken reasonable care to prevent the loss, disclosure or destruction in question.

There is a duty under the Act not to disclose personal data to third parties, which is virtually the same as the common law duty relating to manual records. A fiction has grown up that the law is much stricter about the confidentiality of computer records, but this is definitely not the case, though the confusion shows how important it is to create one law about the holding of personal information in whatever form. The Act specifically exempts from the non-disclosure provisions disclosure required by statute or by order of court, made for the purpose of obtaining legal advice or in the course of legal proceedings, requested or consented to by the data subject or someone acting on his behalf, disclosure to a servant or agent to allow him to perform his functions (a secretary or receptionist, for example), and in any case in which the disclosure is urgently required for preventing injury or damage to the health of any person or where there are reasonable grounds for believing that to be the case. Disclosure of research data is exempt if the individual to whom it relates is not identifiable from the data or other information transferred with it.

There is no doubt that epidemiological research into work-related illness will be greatly assisted by the trend towards computerisation of occupational health records, and as the use of computers becomes more widespread more and more health professionals will need to become acquainted with the Data Protection Act.

3.15   The employer's confidence

It is not only the medical and nursing professions who are obliged to keep secrets: a contract of employment contains implied terms imposing duties of trust and confidence on both employer and employee. These obligations are automatically implied into every employment contract; there is no need to obtain an express undertaking. Many cases have been taken to court in which an organisation has tried to obtain an injunction to prevent its servant (or former servant) from revealing trade secrets to its competitors, or, when the secrets are already out, has sued for damages for the commercial loss caused by his revelations. If the secrets have been used to make large profits, the employer may choose instead to sue for an 'account of profits' to recover all the ill-gotten gains, as did the British Government when a sergeant in the British Army abused his position of trust to assist smugglers in Egypt where he was stationed, making a profit of £20,000 which he was ordered to pay to the Crown, his employer (*Reading* v. *Attorney-General* (1942)). The duty of trust extends to independent contractors who acquire confidential information from their privileged position, like a solicitor to a trust who makes a large profit in a takeover bid of a family company owned by the trust by using his inside knowledge. As with medical confidence, the public interest may justify disclosure. The employee who discovers that his employer is illegally threatening the health of his workers of unlawfully discriminating on the ground of sex or race may report the matter to the appropriate agencies, or even the media, as in *Lion Laboratories* v. *Evans* (above).

How does all this apply to the OH department? It may be that in the course of his employment the doctor or nurse may acquire information which is commercially valuable. He may be consulted about the health risks of a new process, or be given details of a new substance not yet on the market. The competitors of the organisation employing him would be very interested in this information but, needless to say, he should keep it as secret as he does his patients' records. But very occasionally the doctor or nurse may feel that he must speak out without his employer's permission in the interest of the employees or the public outside the factory gates. This is obviously not a step to be lightly taken. Except in an emergency, there should first be consultation with management and with senior health professionals in confidence. Management may be reminded that they have statutory duties under the Health and Safety at Work Act towards both their employees and persons not in their employment. If the doctor or nurse feels that the only way to protect others is to expose a danger this must be done, but he or she should be careful to check the facts and to give the information to those who can best investigate it, probably the HSE, who have a right to inspect the premises at any time, take samples for analysis, etc.

Attempts by health authorities to impose contractual obligations on their staff not to speak to the media about health authority matters led to many protests from professional groups. A 1992 circular from the Secretary of State for Health upheld the ultimate freedom of speech of NHS workers, but the dismissal of Graham Pink, a Stockport nurse who was publicly critical of understaffing in a geriatric ward, and a report published in 1993 which found that half the 31 professionals studied who had made allegations of fraud, malpractice or inadequate care or abuse of patients had either lost their jobs or been put under pressure to resign, give cause for concern. Mr Pink's complaint of unfair dismissal was settled out of court.

3.16 *Official secrets*

Any person who obtains information 'owing to his position as a person who holds or has held office under Her Majesty' is bound not to communicate it to unauthorised persons, on pain of conviction of a criminal offence. The Official Secrets Act binds all Crown servants and anyone who receives information from them knowing or having reasonable cause to believe that it is given in contravention of the Act. The fact that someone has or has not 'signed the Official Secrets Act' is legally irrelevant. Doctors and nurses who are employed by the Crown or who work in organisations with Government contracts should be aware of their responsibilities.

The Official Secrets Act 1989 names six kinds of information as needing protection: defence, security and intelligence, international relations, information received in confidence from other governments, information useful to criminals and the interception of telephone calls and mail. It is a defence that the defendant did not know or have reasonable cause to believe that disclosure would be damaging.

3.17 *Defamation*

An action for breach of confidence is almost always a complaint about the release of a truth which the plaintiff wanted to conceal. Defamation concerns lying statements which have damaged the plaintiff's reputation. A statement proved true cannot give rise to an action in defamation. The burden of proving a statement true is on the defendant.

To sue in defamation, the plaintiff must show that a defamatory statement has been made about him, that is a statement likely to bring him into 'hatred, ridicule or contempt'. If the remark is made in a permanent form (writing, film or broadcast), it is libel, if by the spoken word, slander. An untrue allegation that someone is an alcoholic, a drug addict, a homosexual, mentally ill or suffering from a sexually transmitted disease or AIDS are all defamatory. It would also probably be actionable to allege that a manager was careless of the safety of the workforce. For a statement to be actionable, it must be made to at least one person other than the object of it. A discussion between a doctor and a nurse in a private office during which one says to the other that he is incompetent would not give rise to an action in slander, but if the same remark is made in the presence of a secretary an action could be brought (unless proved to be true!)

The law recognises that sometimes allegations have to be made. Proceedings in Parliament and in a court of law are absolutely privileged; this means that no action can be brought for anything said there. In other situations there is qualified privilege: the statement may be made with impunity as long as the maker honestly believed it to be true and confined it to those who had a moral, legal or social interest in receiving it. A doctor is asked to report to the personnel department on a particular employee, with the employee's consent. He writes in confidence that the man is a liar and a malingerer. No action in defamation is possible as long as the belief was honest, even if the accusations were untrue, but if the doctor was acting from motives of spite he may be liable.

There is no action for privacy in English law, though legislation is likely following public disquiet about Press activities, for example the harassment of close relatives of health professionals disclosed to be suffering from AIDS.

Chapter 4

Pre-employment Screening and Health Surveillance

4.1 The purpose of 'medical assessment'

This chapter is concerned with all forms of 'medical assessment', of prospective employees or those already in employment, which are designed to give an indication of either the effects of work on the individual or the suitability of the individual to do particular work. When I consult a doctor or nurse, it is normally because I am suffering from a particular symptom and need advice on how, if possible, it can be alleviated. The health professional and I are in a therapeutic relationship. But occasionally I ask the doctor or nurse for reassurance that I am in rude health. My apparent hypochondria is the result of my need, or that of my employer or prospective employer or the community at large, to detect any adverse indications before they become a problem to me or to others.

There are many methods of assessment: the administration of a simple questionnaire, an interview with an OH nurse, a selective clinical examination, e.g. of vision, physiological or biological tests and a full clinical examination by a medical practitioner. Using the terminology of specialists in occupational medicine, the following broad categories may be discerned:

(1) The routine medical examination, e.g. pre-employment, or regular routine examinations of vehicle drivers or senior executives.
(2) Special examination of workers identified as exposed to a particular hazard, e.g. audiometric tests, urine tests for workers exposed to carcinogens affecting the urinary tract, blood tests for workers exposed to lead.
(3) Screening tests, i.e. 'sorting through a group or population to identify a relatively small number with a certain characteristic' such as colour blindness, cancer of the cervix, hypertension, diabetes.

(4)   Biological testing, used to identify workers whose perform-
      ance is or may be affected by the consumption of drugs or
      alcohol.

The Gregson Report concluded that the main purpose of an
occupational health (OH) service is to prevent occupational disease
or injury. It has been suggested that there are two kinds of
preventive measure: primary and secondary. Primary prevention
seeks to break the chain between the risk from exposure to a hazard
and its effect on the worker. It includes the complete prohibition of
certain highly toxic substances, engineering controls to reduce
exposure to less hazardous agents, the provision of personal
protective equipment, the monitoring of the environment to ensure
that it complies with established standards, the biological
monitoring of the worker to assess exposure and the exclusion of
workers with a particular susceptibility to the hazard. Secondary
prevention is the detection of harmful effects before the worker
starts to show symptoms; he may then be removed from the danger
before, it is hoped, real harm is done. A third type of preventive
measure, in the form of an epidemiological survey, seeks to identify
hazards by analysing data from a large number of workers, but
once statistical effects cast suspicion on a substance or process
harm has already been done and the prevention of further harm
can only be directed at the workers who come after. Medical
assessment has, therefore, the following goals:

(1)   The early pre-symptomatic detection of disease.
(2)   The evaluation of the effectiveness of engineering controls and
      personal protection.
(3)   The detection of health effects previously unknown and
      unsuspected.
(4)   Suitable job placement.

These scientific aims cannot be criticised; it is the methods
employed and the use made of the information obtained by the
scientists which gives rise to controversy. Kahn-Freund, in his
important work, *Labour and the Law*, wrote: 'Any approach to the
relations between management and labour is fruitless unless the
divergency of their interests is plainly recognised and articulated',
or as an American judge put it: 'The war between the profit-maker
and the wage-earner is always with us'. Management is concerned
with production in the most cost-effective manner. On the whole,
young and healthy workers produce more than the old and
disabled, so there is always an underlying prejudice against the

employment of those whose health is poor. Workers need employment to live and support their families. Many are willing to work in an unhealthy environment as long as they need the money. Men worked in coal mines knowing that they would suffer respiratory problems because there was no other work. The Welfare State provides a cushion for the sick and disabled but it is a thin and uncomfortable one.

The earliest protective legislation was in the field of health and safety because whenever workers entered into bargaining negotiations with employers they were compelled by economics to throw in their physical security as part of the bargain. Employers may consider that medical and biological tests are for the welfare of the employees, but the employees may believe that they are an excuse to make workers redundant. The State explains to workers showing harmful effects that it is for their own good to be excluded from the damaging process, but they may reject this paternalism. Employers argue that they must be able to choose a worker who fits the job they have available. Employees reply that the job should be adjusted to the worker, not the worker to the job. If some workers are particularly susceptible to a substance, why should the solution be to dismiss them and employ a substitute? Why should the law not compel the employer to take precautions to make his process safe for everyone?

Should health professionals accept any kind of assessment, or should they refuse to take part in procedures which may be regarded as an unnecessary invasion of the privacy of the individual? Should they participate in the administration of a questionnaire which asks about life style and sexual preference? Is an employer entitled to demand information about all aspects of an applicant's medical history, and that of his parents and grandparents, as the price of a job?

The OH doctor or nurse is in a very delicate position when it comes to examining the workers. It is important to ensure that they are not perceived as assisting management to reject unwanted staff on spurious medical grounds. For this reason, the ethics of medical assessment are as important as the law. All matters of ethics are open to debate, but it is suggested that the following principles would receive general acceptance:

(1) A programme of assessment should only be undertaken if the results will be of substantial benefit, either to individuals affected by disease or other workers who may be assisted to avoid it. The discovery that asbestos causes cancer did not

benefit those unfortunates who died of lung cancer or mesothelioma but it enabled thousands of others to avoid exposure.
The examination should be designed to detect specific effects.
Routine examinations of general health may bring a false
sense of security and take time and money that could be better
used.

(2) The methods used must be reliable. Too many false positives
or negatives make any form of testing unacceptable. Tests
must be both sensitive (giving positive results in the presence
of disease) and specific (giving negative results in the absence
of disease). Sensitivity is more important than specificity,
because positive results will be investigated but negative
results will not. Audiometry tests must be able to test
permanent hearing loss, not just whether a worker is suffering
from a bad cold, so that the numbers and timing of tests must
be scientifically determined. American doctors have pro-
pounded the theory that routine examinations for cancer of
the bladder, which takes years to develop, should not as at
present in the UK be given to workers in the early years but
only after a substantial period of exposure, otherwise the
statistics will falsely reassure that workers are not being
exposed to dangerous levels.

(3) The methods of assessment must not in themselves create a
hazard to the worker which outweighs their preventive value.
One million low back radiography tests are done every year in
the USA, although their effectiveness has not been established
and they are estimated to result in sixteen leukaemia cases
every year through radiation exposure.

(4) Routine examinations should not be used as an alternative to
engineering controls or environmental monitoring but in
addition to them.

(5) Personnel must be adequately trained to conduct and interpret the results of physiological or biological tests.

(6) Individuals must consent to be examined or tested and must
be told the results of any tests which should not normally be
released to third parties without the subject's consent.

The Health and Safety Executive in 1990 published *Surveillance of
People Exposed to Health Risks at Work*, the purpose of which is to
help all those responsible for and concerned with the control of
occupational health risks to decide when surveillance of those
exposed to such risks should be introduced and what form it
should take.

4.2   Pre-employment medical screening

The Health and Safety Executive (HSE) (*Guidance Note MS 20* (1982)) considers that routine pre-employment health screening of general health in an unselective manner is a waste of time and money, other than a self-administered questionnaire which can be examined by a doctor or nurse. It recommends that a more complex procedure be undertaken only in the following circumstances:

(1)   the job requires the worker to enter a hazardous environment to which he has not previously been exposed (e.g. compressed air, deep sea diving, ionising radiations, lead);

(2)   the work contains specific hazards to the community at large (e.g. transport, health care or catering);

(3)   the work demands high standards of physical or mental fitness (e.g. police, fire services);

(4)   industries where there is a statutory obligation to examine workers before they commence employment (e.g. work with compressed air, work involving exposure to ionising radiations and lead).

Pre-employment examinations involve an assessment of (i) the applicant and (ii) the job. As well as specifically job-related defects (previous high exposure to dangerous substance, allergy to substance, genetic disposition to adverse reaction etc.), many employers are reluctant to employ an applicant if his general health is such that he is likely to be off sick for substantial periods. They accept that established employees may fall ill, but they do not want to take on 'bad risks'. In a time of high unemployment they can be as selective as they wish. There are only a few legal controls.

The OH professional should not be party to the exclusion of job applicants unless there is clear evidence that the job carries specific health risks and also that certain individuals are likely to be more affected than the general population. The HSE gives two examples: candidates for work with lung allergens who have a history of asthma should not be automatically excluded (though an individual medical assessment may be appropriate), but those who wish to work in regular contact with irritants and who have a firm diagnosis of eczema should. If the employer is concerned that the asthmatic whose condition is aggravated by the lung allergen will sue him for negligence, the opinion of experts who can demonstrate that it would be necessary to exclude many job applicants in order to be sure of rejecting the one susceptible individual can be sought. It can then be argued that the employer, and his OH personnel, have taken reasonable care. But it must be said that with high unemployment the employer is free to reject as many candidates as he wishes. Only if he breaches one of the following statutes will any legal liability arise.

4.3   Disabled Persons (Employment) Acts 1944 and 1958

These order employers who employ twenty or more employees to employ three per cent of registered disabled persons. It is a criminal offence to appoint an able-bodied person to a vacancy unless the quota of disabled persons is complete. The Act only affects those who are registered disabled; it is a voluntary decision of the worker to apply to have his name put on the register and receive a green card which is evidence both that he is disabled and that he is fit to undertake employment of a suitable kind. Certain employments (lift operator and car park attendant) have been designated as specially suitable for disabled persons who must be given those jobs. These employees are not included in the three per cent quota. It is generally acknowledged that the legislation is ineffective. There are not enough registered disabled for all employers to meet the quota. An employer who is below quota can obtain a permit from the local jobcentre to fill a vacancy with an able-bodied worker if he can show that there are no suitable registered disabled people available. He can obtain a bulk permit to authorise in advance the engagement of a specified number of workers during a period of up to six months. Bulk permits are issued on the understanding that employers will notify details of vacancies which arise to the local jobcentre, and that they will consider sympathetically the engagement of any suitable registered disabled people who may become available. The law only protects green card holders; many disabled people prefer not to be 'labelled' as such and do not register. Enforcement procedures are lax. Apart from these very limited provisions, there are few laws prohibiting discrimination against the disabled, unlike sex and race discrimination.

Let us look more generally at the job applicant with a worse than average health record. The Manpower Services Commission (MSC) produced a Code of Good Practice on the Employment of Disabled People (1984) which was revised by the Employment Service in 1993. It makes the following important points:

(1)   There is evidence that disabled workers are no less safe than

others. A survey by the British Epilepsy Association found that in companies employing epileptics this particular group had a better than average safety record. In certain cases specific precautions may be needed to ensure that a disabled worker will be safe, as in the event of a fire. Advice can be obtained from EMAS and grants to help with the adaption of premises and equipment are available from the Disablement Advisory Service. Grants to enable an employer to employ a disabled person in a job for a trial period can be obtained through the Disablement Resettlement Officer. Information about these and other schemes can found at any jobcentre.

(2) Disabled workers do not necessarily have more time off work than others, depending on the nature of their disability. Deaf people are not any more or less prone to ill-health than other workers and if they are working in noisy environment they are at less of a risk than other workers. (If there are doubts about the disabled person's ability to cope, the Employment Protection (Consolidation) Act 1978 allows a probation period of up to two years' continuous employment before any employee can complain of unfair dismissal (see Chapter 8).)

(3) The Occupational Pensions Board, in a document published in 1977 reported that pension schemes seldom debar disabled people from participating.

The Code of Practice, which is not legally binding, recommends that no disabled candidate should be rejected on the basis of doubts about safety or fitness, when health checks or a medical examination could dispel such doubts. Part 1 of the Code, for senior managers and directors, outlines the benefits of a constructive approach to employing people with disabilities. It also suggests policy objectives and how to achieve them. Part II is designed to be used as a reference by personnel and line managers and gives practical advice on recruiting disabled workers and integrating them with the rest of the workforce.

By the Companies (Directors' Report (Employment of Disabled Persons)) Regulations 1980, companies which employ on average more than 250 employees must report annually on company policy regarding the employment, training, career development and promotion of disabled people.

English law may be contrasted with that obtaining in the United States. The Americans with Disabilities Act 1990 states that employers may not use the results of a medical examination to exclude a job applicant unless they are such as to make the applicant incapable of adequate job performance even after reasonable provision has been made to accommodate the disability. Attempts to introduce similar protection for the disabled into the laws of this country have so far failed.

4.4   Sex Discrimination Acts 1975 and 1986

Employers are not permitted to discriminate against a job applicant on the ground of sex or, in the case of a married person, of marital status. Assumptions based on sex stereotypes cannot be made about the physical or mental abilities of a candidate. Thus, an employer who turns down a female applicant because the job is a heavy one and 'a woman couldn't cope' will be acting unlawfully. However, if the employer can prove that this particular job requires qualities which this particular applicant cannot fulfil he is not acting unlawfully if he rejects the application. He is then denying the applicant the job on grounds of suitability, not sex. Mrs Thorn applied for a job which involved radial drilling. She was refused it because she was short and slight and might therefore injure herself. The employers were able to convince the tribunal that had the application come from a short slight man they would also have refused him the job (*Thorn* v. *Meggitt Engineering* (1976)). This was not an act of sex discrimination, but a genuine assessment of the candidate's qualifications. If the employer rejects an applicant on such grounds he is in a much better position to defend himself if the applicant has been interviewed and assessed *before* the decision to reject the application (see Chapter 9).

It is common for employers to ask female job applicants to give answers to a pre-employment questionnaire which asks them about specifically female problems such as painful periods or premenstrual syndrome. Should the (probably rare) applicant who honestly admits that she regularly has to take one or two days off at the time of her period be excluded? The justification for this apparent sex discrimination would be that a male applicant who admitted to such regular absences for a health reason would also be classified as unfit. OH personnel are advised that this is a sensitive issue which may give rise to resentment and mistrust and that the question should probably only be asked if it is made clear that the purpose is to assist employees with a health problem, not discrimination. This is especially the case where the job is one predominantly done by males, because it may be seen as an excuse to keep women out.

The HSE (Surveillance of people exposed to health risks at work

(1990)) recommends that the more stringent statutory limits on exposure and absorption for women of reproductive capacity because of risks to the foetus (for instance from lead or ionising radiations) should not be used as a basis for discrimination against the recruitment of women. This official guidance may persuade a tribunal that such discrimination is unjustifiable and therefore unlawful.

A difficult question which has yet to be answered by the judges concerns employers who demand compulsory blood tests for all prospective employees with the intention of rejecting anyone who is carrying the AIDS virus (HIV). Because, at present, most of those affected are males, such testing is a form of indirect discrimination against men. The employer will therefore have to justify it by showing that there is a good reason for exclusion other than that of sex. Possible justification might be the danger to others. British Airways has already decided to screen all candidates for pilots' jobs for HIV antibodies, lest loss of memory and intellectual function create a risk to passengers, but in most jobs there is no such excuse. I would argue that the prejudice of other workers is insufficient justification, because unlawful discrimination is not excused by the equal prejudice of others. If the job is one in which the employer has to invest in the worker by giving him expensive training, the prediction, more and more likely as we learn about the virus, that he will not be able to give any length of service could be economic justification. Pension considerations (unless accompanied by immediate life assurance) are unlikely to be relevant, since someone who is HIV positive will probably not work long enough to qualify for a pension. Where an employee is to be sent to a country like Saudi Arabia, which demands an HIV test as part of immigration control, that also would justify testing. Doctors and nurses who carry out HIV testing have an ethical duty to provide pre-test counselling, and to make provision for further counselling should the test prove positive.

Employers might also be reminded of the 'window', the period between exposure and the appearance of antibodies in the blood during which the test would show a false negative, and of the ever present possibility that someone who tested negative yesterday may be in contact with the virus tomorrow.

4.5 Race Relations Act 1976

This prohibits discrimination on grounds of colour, race or ethnic or national origins. It is, however, permissible to discriminate indirectly if it can be shown that there is a good reason other than race (see Chapter 9). Care must be taken with the questions asked of job applicants. An occupational physician in the West Midlands who asked negro candidates whether they used cannabis, but did not ask the same question of whites and Asians, was accused of racial discrimination. It is permissible to ask applicants if they drink or smoke or take drugs, but all applicants should be asked the same questions. Note that you can be held to have discriminated unlawfully if the *effect* of what you do is discriminatory, whatever your *motive* may have been.

There are special legal rules about nationals of other EC countries. Every citizen of any Member State in principle has the same right to employment on the same terms as a national of the 'home state', assuming that he has the necessary qualifications. The same applies to social security provision.

Scientists are beginning to claim that they can predict which members of the population will be more susceptible to certain substances by genetic screening. For example, sickle cell trait increases susceptibility to hypoxia and alpha-1 antitrypsin deficiency predisposes to the development of emphysema in some circumstances, as after exposure to cadmium fume. Molecular biology may soon enable us to detect sensitivity to occupational carcinogens. If inherited traits are found more often in one racial group, the workers may conclude that these techniques are a disguised form of racial discrimination. The law permits the use of this kind of test in selection for employment as long as the employer can prove that it is justified by the need to protect the workers. He would have to show that the test was reliable and the risk of serious harm to susceptible workers substantial. The BMA (*Medical Ethics Today* (1993)) considers that genetic screening should be optional. 'The purpose of the test should not be to exclude people from employment who are considered by the company to be an economic risk, or to avoid the implementation of safer working conditions or practices which would be of benefit to all employees. Furthermore, employees or prospective employees must have the right to refuse genetic screening without prejudice to their employment prospects.'

4.6 The Rehabilitation of Offenders Act 1974

The Act provides that, if a person convicted of a criminal offence completes a specified period (which varies with the sentence imposed – ten years for a sentence of imprisonment of more than six months but not exceeding thirty months, seven years for six months or less etc.) without reoffending, he can, in effect, forget the conviction, which is described as spent, when he is applying for a job. Most of the 'caring' professions – doctor, nurse, social worker, teacher – have been excluded, as well as the administration of justice, betting, insurance, finance and work with explosives. A conviction which has become spent or any failure to disclose it is not a proper ground for excluding a person from any office, profession, occupation or employment. The Act does not provide for any remedy for the person excluded, but the courts might be willing to create an action for damage for breach of statutory duty. It would be an unfair dismissal to sack an employee because he had kept a spent conviction from his employer.

4.7 Does the job applicant have to answer the employer's questions?

Apart from statutory restrictions, the employer is virtually free to require any questions or examinations he likes. Since there is no legal obligation on a job applicant to volunteer information about himself, unlike an applicant for insurance, the employer is well advised to ask specific questions if the matter is material to him. The job applicant is free to refuse to answer, a hollow freedom when he does not get the job. On the other hand, managers have obligations to employees and to the public as well as shareholders: it is understandable that they are unwilling to take risks.

This dilemma is well illustrated by the case of *O'Brien* v. *Prudential Assurance Co.* (1979). O'Brien applied for a job with the Prudential as a district agent visiting people in their homes. He had a history of mental illness, including periods in hospital. However, at the time of his application, he had not had to consult a doctor for other than minor complaints for four years. When he applied for the job he was required to fill in a form which asked whether he had ever suffered from a serious illness and was also medically examined by a medical referee who asked whether he had ever consulted a psychiatrist or suffered from nervous or mental disorder. O'Brien deliberately lied in answer to these questions because he knew that he would not get the job if he told the truth (this was later confirmed by the employers). Having obtained employment, he proved to be completely satisfactory and colleagues and superiors spoke highly of him, but in the following year he applied for life assurance with the company and gave consent for his GP to give a medical report. Thus, the whole of his previous medical history was disclosed. The junior manager recommended to his senior that a report be obtained from a consultant psychiatrist, but the senior manager instead referred the case to the Company's Principal Medical Adviser. He recommended that O'Brien be dismissed and this was done without further medical evidence being obtained. The reason given was the previous medical history which had not been disclosed in the job application.

Consultant psychiatrists gave evidence at the industrial tribunal where O'Brien was complaining of unfair dismissal that there was no reason why he was currently unsuitable for employment and it was argued that a consultant's report to that effect should have been solicited by management before reaching a decision. The tribunal determined that the reason for the dismissal was not his conduct or capability in employment but his past medical history. They held that employers are entitled to lay down qualifications for candidates and to specify as a policy that they would not employ someone to visit customers in their private homes who had a history of mental illness. Medical evidence from consultants would not have made any difference to the decision, because it was based on the past not the present state of O'Brien's mental health.

When the case went on appeal to the Employment Appeal Tribunal it was decided that past medical history was in a sense a qualification for a job, so that an applicant with a history of ill-health could be regarded as lacking an essential requirement. However, 'it must depend on the facts of the case as to whether for a particular employment the employer behaves reasonably in stipulating the conditions which he seeks to establish'. Was it reasonable for the Prudential to exclude someone from this particular job who had been in mental hospital? The judge said that it was reasonable not only to impose such a condition at the beginning but also to enforce it in the case of someone who had misled them as to facts which they regarded as important:

'It is necessary to stress that this practice or policy is, it appears, one limited to appointment as a district agent, with the particular features of the employment to which we have referred. What the position would be in relation to some other category of employment, where a person had a history of mental illness, is a different question and would always need to be looked at on the particular facts of the case.'

The advice of the Health and Safety Executive (*Guidance Note MS 20* (1982)) is that there are some jobs where the absence of a history of certain types of mental illness might be regarded as a legitimate medical criterion for selection for the job, but in most cases this is not so.

As to the failure to obtain medical evidence other than from the GP and the occupational physician, the Appeal Tribunal considered that it would have been better to obtain consultant advice, but that they would accept the tribunal's finding that it would not have made any difference, because neither consultant who gave evidence was able to guarantee that there would be no recurrence of the illness. The House of Lords decision in *Polkey* v. *Dayton Services* (1987), which emphasises the importance of following a fair procedure, probably means that it is now advisable to obtain such a report.

The following lessons may learned from the O'Brien case:

(1) An employer is entitled in law to ask questions regarding previous medical history and require medical examinations of job applicants.

(2) If an employee lies in answer to such questions the employer may, on discovery of the truth, dismiss him, but only if the employer can show that he would not have employed him had he known the truth and that this would have been a reasonable response because of the nature of the disability and the nature of the job.

(3) A person's previous history of ill-health may be held to render him unfit for a job because of the danger of recurrence, despite his good health at the time he applies for it.

(4) In a case like O'Brien's, it is preferable for the occupational physician to obtain a consultant's opinion. There are different forms of mental illness; every year about five million people consult their doctors about emotional or mental problems and only a minority of the people who become mentally ill suffer long-term or permanent disability that makes them difficult to employ.

The case of Beverly Allitt, the nurse suffering from Münchausen's syndrome by proxy, who murdered several children in her charge, has led to calls to exclude anyone with a history of psychiatric illness from the caring professions. It is important that the understandably horrified reaction of the public to the facts of that case does not lead to a witch-hunt. The nursing profession advocates the setting of national selection criteria for those seeking employment as a nurse, a sensible method of ensuring that such criteria are properly debated and scientifically justified. An audit of selection criteria in the NHS has been instituted. Ultimately, it will be in the discretion of the OH professional conducting the confidential pre-employment procedure as to whether he or she classifies the job applicant as unfit (clinical details should not, of course, be revealed without the applicant's consent).

In the USA there is a growing movement towards pre-employment and regular biological screening for alcohol and drugs of abuse, and American companies here are beginning to follow this practice. British Nuclear Fuels is reported to have introduced drugs screening for all its employees. There are no legal objections to the practice at the pre-employment stage as long as it is realised that applicants are free to refuse the test if they wish. The legality of testing those already in employment is considered below. There may, however, be ethical objections. Is the method of testing accurate and reliable? How many cases can it detect if job applicants know about it beforehand? On the other hand, job applicants are very unlikely to give such information in reply to questioning alone. In *Walton* v. *TAC Construction Materials* (1981), a prospective employee was interviewed by an OH nurse and was asked if he suffered from any serious illnesses. He replied that he did not. In reality he was injecting himself with heroin. Soon afterwards, Walton, of his own volition, consulted a GP who treated him and substituted oral diconal for intravenous heroin. The employers found another employee who was taking heroin and this led to the discovery that Walton was a registered drug addict and was being treated for it. Management sought advice from the occupational health physician who stated that had he known that Walton was a drug addict he would not have recommended him for employment and that he could not advise the company to retain a drug addict in employment. The doctor considered a drug addict to be a potential danger to himself and others, that the incidence of relapse was high, and that if he was kept on he would have to be supervised. The employers decided to dismiss without consulting the GP. Note that the decision was not based on the nature of Walton's job or the particular facts of his case but on a general policy (this was a construction company). The tribunal held that the dismissal was fair and the Employment Appeal Tribunal declined to interfere. It was within the range of reasonable responses for an employer to say that the risks attendant on employing a drug addict in his business precluded him from so doing. Latterly, views more sympathetic to the addict have gained support, because this kind of policy deters him from going for treatment. If Walton had never approached his doctor his problem might not have been uncovered and he might still have his habit and his job.

When a prospective employee conceals disabilities from the employer because he fears that otherwise he will not be appointed, he puts himself and others at risk. In one case a man who suffered from epileptic fits kept this fact from the employer. While on a working platform 23 feet above the ground he had an epileptic fit, fell off the platform and was killed. It was held that as the employer could not be expected to know of the employee's disability he was not negligent in allowing him to work at a height (*Cork* v. *Kirby Maclean* (1952)). 'Paradoxically, the question of legal liability for accidents is altered to the disadvantage of the employer in that, having had the worker examined, he is more likely to be held at fault if any accident occurs to which the medical condition of the party may conceivably have contributed. This of itself forces the employer to be conservative in interpreting any medical findings, tending to reject any worker subject to any suspicion. Thus the employer should carefully appraise the need for such procedures in the light of individual cases'. (*HSE Guidance Note MS 20* (1982).

4.8 Examination for entry to the pension scheme

The OH physician may frequently fulfil a dual role, examining the prospective employee both for suitability for the job and as a candidate for a pension. In the first instance he is reporting to the employer and in the second to the trustees of the pension fund. Doctors should not confuse pensions and insurance. Insurance policies cover the risk from the first day, whereas pensions commonly depend on a period of service. Also, the best possible employee from the standpoint of the pension fund is one who drops dead without any surviving dependants on the day of his retirement, not one who draws his pension for twenty years.

4.9 Surveillance of the existing workforce

More legal constraints are imposed on the employer who has accepted the worker into his organisation. For this reason, personnel departments are concentrating increasingly on the pre-employment stage, hoping to weed out the potential problem cases at the outset. Once the agreement has been made, the worker, employee or independent contractor, has the protection of the law of contract, and after two years' continuous service employees can complain to an industrial tribunal that they have been unfairly dismissed (see Chapter 8).

A contract is concluded and becomes legally binding once there is an agreement; there is in general no necessity in law for the agreement to be in writing, though written evidence is better than oral in the event of a dispute. If an employer desires the inclusion in the contract of a particular obligation he is well advised to spell it out clearly in writing. In consequence, an employer who has decided that he wants his employees to undergo medical examinations or tests, either at random or at regular intervals, must specify this as a term of the contract at the time the worker is hired. If this has not been done, it will be a breach of contract for the employer to try to insist on testing at a later date unless he can show that there was an *implied* term to that effect or he can persuade the individual workers to agree to its introduction (trade union officials have no power to give a group consent to a change in the terms of a contract), or there is an overriding statutory duty to undergo testing.

The leading case on this point concerned no less a person than a consultant orthopaedic surgeon, Mr Bliss. Complaints had been made about his alleged intemperate behaviour to the Regional Medical Officer who had referred the matter to a committee of three consultants (the 'Three Wise Men'). They concluded that Bliss was sane and that it was the orthopaedic department which had suffered the breakdown. After this the RMO asked Bliss to undergo a psychiatric examination by a consultant nominated by the health authority. When Bliss refused, he was suspended on full pay from his post as a consultant and denied access to the hospital. The Court of Appeal decided that there was no legal obligation on a consultant to submit to a medical examination when asked, unless the employer had reasonable grounds for believing that he might be suffering from physical or mental disability which might cause harm to patients or adversely affect the quality of the treatment of patients, when there would be an implied term in the contract. There were no reasonable grounds here, because the Wise Men had already investigated the case and found him sane. In consequence, the health authority were in breach of contract and must pay compensation for the earnings from private patients which Bliss had lost, though not for injury to his dignity and feelings, because that is not a recognised head of damages for breach of contract (*Bliss* v. *SE Thames RHA* (1985)). There was no action for defamation because the allegations about mental instability had been confined to those covered by qualified privilege; if a statement had been made to the Press or the matter gossiped about at the golf club, Bliss could have claimed for loss of reputation (see Chapter 3).

Clearly, the courts are ready to accept that in the case of some employees there is an implied term in the contract of employment that they will submit to a medical examination when reasonably requested. Which employees? The answer seems to be those like health professionals whose illness could constitute a risk to others. Terms will not be implied if the reason for the tests is to protect the employees themselves or to ensure that the employer is not sued for negligence. Does this mean that an employer can lawfully introduce HIV or alcohol or drugs testing for all without the consent of his employees? Only if he can show that there is a real risk to others if the problem remains undetected. In the case of HIV this will be confined to a very few jobs.

Fear of AIDS has led to pressure on the Government to introduce compulsory testing of health professionals for HIV. Health professionals have responded with a call for the compulsory testing of patients. Department of Health guidelines and pro-fessional bodies prefer to place the responsibility on the health professional to volunteer for a test if he or she thinks there may be a risk. This has already been discussed in Chapter 2.

Following on from this, the far greater risk of infection from hepatitis B has led to recommendations from the Advisory Group on Hepatitis, published in 1993. They advise that all health care workers who perform exposure-prone procedures (where there is a risk that injury to the worker may result in exposure of the patient's open tissues to the blood of the worker) and all medical, dental, nursing and midwifery students should be immunised against hepatitis B, unless immunity as a result of natural infection or previous exposure has been documented. Their responses to the vaccine should be subsequently checked. Staff whose work involves exposure-prone procedures (surgery, dentistry and renal dialysis are particularly high risk areas) and who fail to respond to the vaccine should be permitted to continue in their work provided they are not e antigen (HBeAg) positive carriers of the virus. Otherwise, employers should make every effort to provide alterna-tive employment. It is recommended that all matters arising from and relating to the employment of HBeAg positive health care workers are co-ordinated through a consultant in occupational medicine.

Since this is only advice, and does not have the force of law, legal

problems may surface where an existing employee refuses immunisation or subsequent monitoring. The employer will have to
argue that there is an implied duty in the health care worker's
contract of employment to agree to these procedures. There is less
difficulty where the employee is entering employment, because his
new contract can and should include an express term. Where an
existing employee tests HBeAg positive, and no satisfactory
alternative post can be found, it may be necessary for the employee
to take early retirement or be dismissed. The employer would
probably succeed in the defence that the employee has become
unfit to do the job, in the same way as if he had succumbed to a
crippling disease. The Faculty of Occupational Medicine recommends that clear procedures should be in place in advance to
deal with cases where the test result indicates a change of duties,
and where the employee or student declines to be tested or
immunised. Hepatitis B is a prescribed disease for health care
workers and benefits are available under either the NHS Injury
Benefits Scheme or the general scheme (for employees in the private sector) (see Chapter 6).

What is the importance of establishing that the employee is in
breach of contract if he refuses the test or the immunisation? It
allows the employer to dismiss him without a breach of contract on
the employer's part, because it is then the employee who has
destroyed the relationship and not the employer.

It also makes it more likely that the dismissal will be upheld as
fair in an industrial tribunal. Sensible employers negotiate with
their workers to include an express term in every contract, pointing
out the benefit to fellow workers as well as to the public. It is not
sufficient, though common, to impose the compulsory medical
only after a period of absence from work; those suffering from
severe psychiatric illness may not take time off, insisting that there
is nothing wrong with them.

Of course, no health professional will wish to be associated with
oppression of the workforce, and the industrial relations implications of medical examinations or biological tests must be taken
into account. In such a sensitive area the role of the OH professional
is as an expert to be able to present the available scientific information to both sides impartially and to stand by the ethical rules of
his or her profession in consultation with other professionals.
Where a positive result in a drugs or alcohol test will lead to disciplinary proceedings, the employee should be given the right to
'appeal' by undergoing a test of the same or a different kind in
order to check that the first test result was accurate. In *Sutherland* v.

Sonat Offshore (UK) Inc. (1993), the employee returned to the mainland after two weeks on an oil rig. A drug test showed the presence of cannabis in his urine. The result was confirmed in further tests, and the employee was dismissed by letter. The tests were conducted by Offshore Medical Services Ltd, a specialist contractor. It was held that the employee had, as it were, been caught red-handed. The employer was entitled to take the view that to hold a formal disciplinary hearing would serve no useful purpose. The dismissal was fair.

The Faculty of Occupational Medicine advises that it is neither necessary nor appropriate for medical or nursing staff to be involved in procedures such as breath analysis for alcohol where these are undertaken as part of a company programme. 'The involvement of health staff in "policing" procedures may undermine their role as confidential medical advisers and any possible confusion should be avoided.' Where OH staff are unavoidably involved they should not give medical advice, because they should make it clear that they are acting as the employer's agent. Scrupulous care must be taken in handling samples, transmission to the laboratory and handling of the results.

The Transport and Works Act 1992 has introduced compulsory testing by the police (including the railway police) of railway employees in a 'safety-sensitive' position. The new provisions apply to drivers, guards, conductors, signalmen, maintenance staff, supervisors of maintenance staff and look-outs. Where a constable in uniform has reasonable cause to suspect that a person is or has been working with alcohol in his body and still has alcohol in his body he may require that person to undergo a breath test. There is a similar power where there has been an accident or dangerous incident and the act or omission of the person tested may have been the cause of the accident or incident. A constable may also require a person suspected of working on a railway when he is unfit to give a specimen of blood or urine if a doctor advises him that the condition of the suspect might be due to a drug. Failure to provide a specimen without reasonable excuse is an offence. These criminal provisions do not allow tests of employees before they start their shift: the employer will have to impose an obligation to undergo such tests in the contracts of employment, if he wishes to do so.

Since the 1992 Act came into force, British Rail has introduced random and 'for cause' urine testing for all employees. The samples are taken and despatched for analysis by the occupational health nurse. Experience to date is that the new system has achieved the acceptance of most of the workers who see it as a means of protecting themselves, and who have confidence in the occupational health staff.

All examinations and tests may only be performed with the consent of the individual: even the police when investigating a serious crime have no power to take blood or breath or other intimate samples without consent, though a refusal by the suspect can be used in evidence against him (s. 62 Police and Criminal Evidence Act 1984). However, an employee who is contractually bound to undergo an examination is not in a strong legal position if he is disciplined or dismissed for refusing. It is a moot point whether an employee can be prosecuted under s. 7 HSWA, which imposes on him an obligation to co-operate with the employer in performing the employer's statutory duties, for refusing to submit to a test where there is no specific statutory or contractual duty on him to do so. I think that our courts would be very unwilling to make a worker criminally liable for an omission to do something for which neither his contract nor Parliament has provided. The section was principally intended to impose on the employee a duty to wear personal protection and make use of safety precautions provided by the employer.

4.10   *Levels of routine health surveillance*

Occupational health risks vary from industry to industry and job to job. Employers must be flexible in both the levels of health surveillance and the personnel employed to carry it out. For example, although a doctor might be in overall charge of biological or physiological tests, the tests themselves can be performed by nurses or other trained personnel. A doctor is not needed if hazards are low and there are easily recognisable and specific symptoms, e.g. a rash, as long as workers manifesting those symptoms are immediately sent for medical advice. In the case of greater risk, health professionals will be needed. The following levels are possible:

(1)   Screening tests. These are designed to identify specific biological or physiological changes in workers known to be at risk. They include blood and urine tests, lung function and audiometry. They should be carried out by personnel specially trained in standard procedures and the results compared with approved reference levels. OH nurses, occupational hygienists and technicians are suitable to undertake

testing. Where tests are required by statute the relevant Regulations and Approved Codes of Practice will have to be followed.

(2) Review of records. This can be employed in cases where there is a known risk but no agreed method of monitoring its effects on workers, e.g. substances creating mood changes. A member of the occupational health team can compare records of sickness absence with environmental records showing degrees of exposure, hours worked and so on.

(3) Clinical examination by a doctor, including both initial and periodic examinations and probably supplemented by tests. These can take place at regular intervals, e.g. airline pilots, or if the results of tests or a review of records so indicate.

4.11   Legal duty to provide health surveillance

Employers can be held civilly liable at common law for failing to introduce health surveillance even in the absence of specific statutory regulation. In *Wright* v. *Dunlop Rubber Co.* (1972) the employers used an anti-oxidant called Nonox S in their process. ICI, the manufacturers, then discovered that it contained free beta-naphthylamine, a known carcinogen and now a prohibited substance, and withdrew it. In 1960 a circular from the Rubber Manufacturers Employers' Association warned that all employees who had been exposed should be screened and tested for bladder cancer, a disease with a long latency period which can be successfully treated if caught in the early stage. Dunlop stopped using Nonox in 1949 but did not introduce urine tests for workers who had been exposed until 1965. Only then was it discovered that the plaintiff had cancer of the bladder (hundreds of other cases subsequently came to light). It was held that in addition to the liability of the manufacturer the employer was liable in negligence for failing to institute tests quickly enough. This should have been done in 1960. At that time there was no statutory obligation (it was later introduced by the COSHH Regulations 1988).

There is now a specific statutory provision applying to all employees other than those working on sea-going ships. This is Reg. 5 of the Management of Health and Safety at Work Regulations 1992: 'Every employer shall ensure that his employees are provided with such health surveillance as is appropriate having regard to the risks to their health and safety which are identified by the assessment'.

Where a risk has been well documented and established, eventually the State may intervene and demand that the employer introduce a health surveillance programme regulated by inspectors with statutory powers. The provisions of such legislation will, of course, vary with each set of regulations, but common factors can be discerned. If a specific problem arises, reference must be made to the relevant regulations, but the following general comments may be helpful:

(1) Health surveillance is either a response to a risk to the health of workers created by continued exposure to a dangerous substance, e.g. lead or ionising radiation, or it is a means of ensuring that an individual who may constitute a risk to the public, e.g. a public service driver, is fit to do the job. The protection of the public is often achieved by a system of licensing under which an applicant must pass a medical test to be given a licence or have it renewed.

(2) In both cases it is usually necessary to examine the worker at the hiring stage so that a 'baseline' can be established. Sometimes, pre-employment screening may reveal defects which lead to the rejection of the job applicant.

(3) The statutory obligation is usually placed on both employer and employee. For example, the Control of Lead at Work Regulations 1980 provide that an employer shall secure that each of his employees employed on work which exposes that employee to lead is under medical surveillance by an Employment Medical Adviser (EMA) or Appointed Doctor (AD) if either the exposure to lead is significant or the doctor has certified that the employee must be under medical surveillance. It goes on to state that every employee who is exposed to lead at work shall, when required by the employer, present himself, during his normal working hours, for such medical examination or biological tests as may be required to allow the employer to carry out his statutory duty. There is, therefore, no need to include a contractual term, because the employee is in breach of the statute and in theory can be prosecuted if he declines to be tested. The Ionising Radiations Regulations 1985 oblige the employee not only to present himself for a medical examination and tests but also to furnish the EMA or AD with such information concerning his health as the doctor may reasonably require. An employer would clearly be justified in sacking a worker who refused these statutory tests, because he would be in contravention of the criminal law (see Chapter 8).

(4) The EMA or AD commonly has a statutory power in effect to stop the worker working by certifying him as unfit if the results of the medical surveillance warrant it. Employers are responsible for taking the necessary action to remove the employee from exposure and remedy an unsatisfactory working environment. Sometimes this will only lead to a brief suspension by the employer, but if the employee is in danger of losing his livelihood, statutory regulations may give the worker a right of appeal against the medical opinion. Regulation 16 of the Ionising Radiations Regulations, for example, allows an employee aggrieved by a decision recorded in the health record by an EMA or AD to apply for the decision to be reviewed. The application must be made to the HSE within three months.

(5) There is little point in regular examination if no proper record of the result is kept. Efficient record-keeping is a vital component of medical surveillance, statutory or voluntary. It is also important evidence for the epidemiologist. The Employment Medical Advisory Service (EMAS) must be notified of the statutory examinations which have been carried out and of those workers certified unfit.

4.12   *The Control of Substances Hazardous to Health (COSHH) Regulations*

The COSHH Regulations are designed to protect all workers whose job brings them into contact with a substance hazardous to health, whatever the nature of the job or the nature of the substance. The duty to provide adequate ventilation, personal protection, hygiene, health surveillance etc. is no longer confined to specified places of work like factories, or specified substances like asbestos.

The need for the comprehensive approach is supported by statistics. The DSS reports a large number of claims for social security in respect of diseases attributable to occupational exposure to substances. As the figures relate only to clearly established cases of prescribed diseases, they are acknowledged to be incomplete. Many people suffer from chronic bronchitis and asthma but no-one knows how many cases are occupationally linked. There were 130,000 cancer deaths in Britain in 1980; it is thought that between two and eight per cent could be occupationally linked (Doll and Peto (1982).

In a Consultative Document issued by the Health and Safety Commission (HSC) in 1984 the objectives of the new laws were set down as:

(1)　to provide one set of regulations covering hazardous substances, including substances not yet covered by specific regulations, and all exposed employees whether or not they work in a factory;
(2)　to set out in the regulations principles of occupational medicine and hygiene;
(3)　to make provision for any future discoveries which may lead to changes in standards of control and improved techniques for the control of exposure;
(4)　to be able to implement the EC Directive on the protection of workers from risks related to exposure to chemical, physical and biological agents at work (1980) and future Directives.
(5)　to enable the Government to ratify the ILO Convention No. 139 on carcinogenic substances.
(6)　to simplify and update the law.

The core of COSHH Regulations is a duty on every employer to carry out an assessment before starting work which may involve exposure to a substance hazardous to health. This is intended to identify the nature and degree of risk. After the risk has been assessed the employer must then determine the precautions needed to protect the workers. There are basically three types of prevention: control of exposure, using engineering controls or personal protective equipment, monitoring of the environment to check that control limits are not being exceeded and health surveillance to detect the early signs of disease and assess the efficiency of the control system. The first is the province of the engineer, the second of the hygienist and the third of the health professional.

'Substance' is defined by the regulations as 'any natural or artificial substance, whether in solid or liquid form or in the form of a gas or vapour (including micro-organisms)'. 'Micro-organism' includes any microscopic biological entity which is capable of replication. A substance hazardous to health means any substance (including any preparation) which is:

(1)　any substance listed in the Classification, Packaging and Labelling Regulations 1984 for which the classification is specified as very toxic, toxic, harmful, corrosive, irritant, carcinogenic, mutagenic or teratogenic;
(2)　any substance for which the maximum exposure limit is specified in Schedule 1 of the regulations or for which the HSC has approved an occupational exposure standard;

(3)   a micro-organism which creates a hazard to the health of any person;

(4)   dust of any kind, when present at a substantial concentration in air;

(5)   a substance, other than those already included, which creates a hazard to the health of any person which is comparable with the hazards created by those other substances.

The COSHH Code advises that a substance should be regarded as hazardous to health if it is hazardous in the form in which it occurs in the work activity, whether or not its mode of causing injury to health is known, and whether or not the active constituent has been identified. Also included are mixtures of compounds, micro-organisms, allergens, etc.

The regulations do not apply to substances already covered by the Control of Asbestos at Work Regulations 1987 and the Control of Lead at Work Regulations 1980. They do not extend to exposure to a substance hazardous only because of its radio-active, explosive or flammable properties, or solely because it is at a high or low temperature or at a high pressure. If a substance is administered to a person in the course of his medical or dental examination or treatment (including research) performed under the direction of a registered medical or dental practitioner, the regulations do not apply. They do not apply to work below ground in a mine.

COSHH applies to exposure which is related to work, not that which is incidental. Regulation 2 states that any reference to an employee being exposed is a reference to the exposure of that employee to a hazardous substance *arising out of or in connection with work which is under the control of the employer*. Thus, the exposure of a nurse to hepatitis B infection while nursing a patient is within the regulations, but not the possibility that a clerical worker might be infected through social contact with other employees. Cigarette smoke will not be within the regulations (though probably within the duty under section 2 of the HSWA), because it is not produced by the work being done, but is incidental to it (see Chapter 5). An employee who is sent to another organisation to work under the control of the latter will have to rely on the controlling employer's duty.

Certain substances are prohibited from use. These are listed in Schedule 2 and include 2-naphthylamine, benzidine, 4-aminodiphenyl, 4-nitro-diphenyl, their salts and any substance containing any of these compounds in a total concentration exceeding one per cent. All these substances, and matches made with white phosphorus, are prohibited from importation into the UK.

Regulation 3 lays down the scope of the employer's duty. The employer is always responsible to those directly employed, his employees, but he is in general also under a like duty, so far as is reasonably practicable, to any other person, whether at work or not, who may be affected by work carried on by him (i.e. under his control). This will extend the employer's duty of care under the regulations to subcontractors and members of the public, both on and off the employer's premises (see Chapter 5). Not all the statutory obligations are owed to this wider group. In particular, the health surveillance provisions under Regulation 11: the employer is only obliged to provide suitable health surveillance for his own *employees*. Regulation 11 states that where it is appropriate for the protection of the health of employees who are, or are liable to be, exposed to a substance hazardous to health, the employer *shall* ensure that such employees are under suitable health surveillance (a term which includes biological monitoring). Contractors must keep their own health records for their employees, and self-employed persons working on site are not covered by this part of the new law, though the protection of their health is within its scope. Also, there will be no statutory obligation to keep a health record of exposure to hazardous substances where that exposure has occurred in the course of work not under the employer's control.

When is health surveillance appropriate? When *either* the employee is exposed to a substance and engaged in a process specified in Schedule 5, unless the exposure is not significant, or 'the exposure of the employee to a substance hazardous to health is such that an identifiable disease or adverse health effect may be related to the exposure, there is a reasonable likelihood that the disease or effect may occur under the particular conditions of his work and there are valid techniques for detecting signs of the disease or the effect'. It has been decided as a policy matter that the degree of significant exposure will not be automatically tied to occupational exposure standards or maximum exposure levels set by the legislation. Each employer will have to determine significance, acting on expert advice.

The following substances and processes are listed in Schedule 5: vinyl chloride monomer (manufacture, production, reclamation, storage, discharge, transport, use or polymerisation); nitro or amino derivatives of phenol and of benzene or its homologues (manufacture of nitro or amino derivatives of phenol and of benzene or its homologues and the making of explosives with the use of any of these substances); potassium or sodium chromate or dichromate (in manufacture); 1-napthylamine and its salts, ortho-tolidine and its salts, dianisidine and its salts, dichlorbenzidine and its salts (in manufacture, formation or use of these substances); auramine, magenta (in manufacture); carbon disulphide, disulphur dichloride, benzene, including benzol, carbon tetrachloride, trichlorethylene (processes in which these substances are used, or given off as vapour, in the manufacture of india rubber or of articles or goods made wholly or partially of india rubber); pitch (in the manufacture of blocks of fuel consisting of coal, coal dust, coke or slurry with pitch as a binding substance). Other substances may be added later by statutory instrument.

The purposes of health surveillance are set out in the COSHH Approved Code of Practice as:

(1) the protection of the health of individual employees by the detection at as early a stage as possible of adverse changes which may be attributed to exposure to hazardous substances;
(2) assisting in the evaluation of measures taken to control exposure;
(3) the collection, maintenance and use of data for the detection and evaluation of hazards to health;
(4) assessing in relation to specific work activities involving hazardous micro-organisms, the immunological status of employees.

What is suitable health surveillance? As an absolute minimum, where the regulations apply, the employer must keep an individual health record, containing the particulars set out in the Approved Code of Practice, including personnel details, a historical record of jobs involving exposure to substances requiring health surveillance and dates and conclusions of all other health surveillance procedures. These records or a copy thereof must be retained for at least 40 years from the date of the last entry. If the employer ceases to trade he must offer them to the HSE. The employer must allow each employee to see his own health record if he asks for it. It is clear that the intention behind the regulations is that the health record is to be kept separately from confidential clinical records. It must be held in a form capable of being linked with the records of the monitoring of exposure, and thus in practice it will have to be made available to hygienists and others outside the occupational health department. It can be foreseen that this may cause conflict, because health professionals will be reluctant to give detailed results of, for example, biological tests, which they regard as clinical data, to others without the written consent of the workers concerned. The Code of Practice advises that clinical data should not be included in the health record, in which only the *conclusions* of all health surveillance procedures and the date on which and by whom they were carried out should appear. These should be expressed in terms of the employee's fitness for work and, where appropriate, will record the decisions of the doctor, nurse or responsible person, *but not confidential clinical data*. The health professional or other responsible person should explain to employees undergoing health surveillance that he will have to notify their employer of his conclusions as to their fitness to work, for entry in the health record.

The Approved Code of Practice contains more detailed recommendations. It envisages at least seven levels of surveillance:

(1) The keeping of an individual health record. This is necessary for *all* employees exposed to a substance hazardous to health.
(2) Biological monitoring, i.e. the measurement and assessment of workplace agents or their metabolites either in tissues, secreta, excreta or expired air, or any combination of these in exposed workers.
(3) Biological effect monitoring, i.e. the measurement and assessment of early biological effects in exposed workers.
(4) Inspection by a 'responsible person' (e.g. a supervisor or manager). A responsible person is someone appointed by the employer, trained to know what to look for, and charged with reporting to the employer. An example given in the Code is a regular inspection by a supervisor of the skin of electroplating workers using chromium compounds.
(5) Enquiries about symptoms, inspection or examination by a 'suitably qualified person' (a qualified OH nurse is given as an example but it is not made clear what other qualification is suitable). An example might be regular inspections of skin by an OH nurse of those working with known or suspected skin carcinogens (e.g. arsenic, pitch, coal tar, shale oil).
(6) Medical surveillance by a doctor, who must be an EMA or an AD where the employee has been exposed to a substance listed in Schedule 5, or any registered medical practitioner in any other case. This may include clinical examinations and measurements of physiological and psychological effects of exposure to hazardous substances in the workplace, as indicated by alterations in body function or constituents.

(7) A review of records and occupational history during and after exposure should be used by management to check the correctness of their assessment of the risk.

'The aim should be to evolve health surveillance procedures which are safe, easy to perform, non-invasive and acceptable to employees.' Judgments as to the likelihood of an adverse health effect must be related to the nature and degree of exposure. Valid techniques of high sensitivity and specificity must be used. It may be necessary to establish normal values and action levels. The results of health surveillance should lead to some action which will be of benefit to the health of employees. Methods of surveillance, options and criteria for action should be established *before* undertaking a programme. The procedures are not mutually exclusive. The results of biological monitoring may indicate the need for a medical examination. The keeping of records is often the only form of surveillance required for those working with carcinogens other than those specified in Schedule 5 and skin carcinogens. This is because of the usually long latent period between exposure to a carcinogenic substance and any health effect. This means that there is no specific statutory obligation on the employer to examine those workers in any way or to employ any personnel to monitor their health. The general duties under the Health and Safety at Work Act and the common law, of course, oblige him to do that which is reasonable in all the circumstances.

Employees exposed to a Schedule 5 substance and process must be regularly under medical surveillance supervised by an EMA or AD at least every 12 months, or more frequently if the doctor requires it. Medical surveillance includes clinical examinations and measurements of the physiological and psychological effects of exposure in the workplace; the exact nature of the examination is at the doctor's discretion. However, as has already been stated, the obligation on the employer to undertake and on the employee to undergo health surveillance is not confined to Schedule 5 substances. The employee who works in *any* job where it is appropriate within the regulations to perform tests or examinations on him has a duty to present himself during working hours for such health surveillance procedures as may be required by the employer. In addition, an employee subject to medical surveillance in connection with a Schedule 5 substance and process must give the doctor information about his health if reasonably required to do so. Since these obligations are imposed by statute, the employee will not be able to refuse to undergo health surveillance without being in

breach of his statutory duty and therefore open to criminal prosecution and disciplinary action by the employer.

The EMA or AD has power to certify in the health record that in
his professional opinion the employee is unfit to work with the
Schedule 5 substance, or to place conditions on his continued
employment. An employee suspended by the EMA or the AD will
be entitled to medical suspension pay (Chapter 8). The employer
will be committing an offence if he ignores the doctor's finding.
Also, the doctor will be able to certify in the health record that the
employee must continue under surveillance after his exposure to
the dangerous substance has ceased. The employer will be committing an offence if he fails to provide for this as long as the
employee continues in his employment. The employee has a right
to appeal against the doctor's decision to suspend him; he should
write to the HSE within 28 days of being notified of it. The doctor
must be allowed to inspect the workplace or the health records if he
requires it.

Although the regulations make detailed provision for regular
medical examination of those employees exposed to a substance
listed in Schedule 5, they are less specific about other substances.
Health surveillance, including medical surveillance, may be
required, but the necessity for medical surveillance by an EMA or
AD only attaches to Schedule 5 cases. Also, the statutory right of
appeal to the HSE against medical suspension outlined in the
preceding paragraph and the right to medical suspension pay from
the employer for up to 26 weeks are restricted to a decision of an
EMA or AD. The position of the OH professional undertaking
routine surveillance for a non-Schedule 5 substance is unclear.
Suppose an employee refuses to allow him to report that he is
unfit? Would it be a breach of confidence to make a report without
consent? Since the employee has a statutory duty to submit to
health surveillance, it is probably the law that the doctor or nurse
can report on simple fitness or unfitness for work (but not the
clinical details) without consent. Employers who voluntarily
provide some compensation for employees suspended or
dismissed in these circumstances are likely to achieve greater co-
operation from the workforce, and this is recommended by the HSE
(*Surveillance of people exposed to health risks at work* (1990)).

4.13 Management of Health and Safety at Work Regulations 1992 and ACOP

Regulation 5 extends the obligation to provide health surveillance
to all employees, other than those employed on sea-going ships. It

is such surveillance as is appropriate having regard to the risks to the employees' health and safety identified in the employer's assessment. The Approved Code of Practice adopts the following criteria:

(1) there is an identifiable disease or adverse health condition related to the work concerned;
(2) valid techniques are available to detect indications of the disease or condition;
(3) there is a reasonable likelihood that the disease or condition may occur under the particular conditions of work; and
(4) surveillance is likely to further the protection of the health of the employees to be covered.

The Code of Practice indicates that the primary benefit of surveillance is to detect adverse health effects at an early stage, but that secondary gains are checking the effectiveness of control measures, providing feedback on the accuracy of the risk assessment and identifying and protecting individuals at increased risk. The forms of surveillance are those already set out in the COSHH Approved Code of Practice and discussed in the previous section of this chapter. The Management Approved Code of Practice places emphasis on the participation of the workforce. The employees concerned should be given an opportunity to comment on the proposed frequency of health surveillance procedures and should have access to an appropriately qualified practitioner for advice on surveillance.

This regulation has the potential for the introduction into UK law by the back door of a mandatory obligation on all employers to provide an occupational health service of some kind. Everything depends on how it is interpreted in practice. The Management Regulations themselves impose only criminal sanctions. Though no civil action for breach of statutory duty can be brought directly for breach of the Management Regulations, it is likely that the civil courts will refer to the Regulations when deciding whether the employer accused of common law negligence has taken reasonable care. Thus, an employer who does not provide health surveillance as required by the Management Regulations has arguably failed to take reasonable care for the health of his employees and would be liable in negligence.

Regulation 5 imposes no duty on employees to submit to health surveillance procedures, as does Reg. 11 of the COSHH Regulations. Difficulty may arise if workers perceive surveillance as a method of excluding the less fit for the benefit of management rather than the protection of the employees. The primary duty on the employer is to care for all his workers, not to try to achieve a workforce of supermen and superwomen.

Chapter 5

Health and Safety at Work: the Criminal Law

5.1 Criminal sanctions

The movement of workers from their rural homes to factories in industrial towns had by the beginning of the nineteenth century created such horrific conditions as to agitate reformers like the Earl of Shaftesbury to promote legislation to force manufacturers to improve conditions at work. The first such measure was the Health and Morals of Apprentices Act of 1802, but the first statute to set up an inspectorate to enforce the new laws was the Factory Act 1833.

From the beginning, the law concentrated on establishing and supporting criminal sanctions. The purpose of the legislation was to deter employers from creating unhealthy and dangerous working practices. The development of a legal right to money compensation for those killed and injured followed only 60 years later. The theory that criminal penalties are a more effective deterrent than the duty to compensate has often been debated. Where the defendant to criminal proceedings is a corporation, it cannot be sent to prison. Fines for health and safety offences are low, especially in the magistrates' courts. The maximum fine available to the magistrates was raised from £2,000 to £20,000 in 1992. The Crown Court has never had an upper limit on the amount of fines. The highest ever fine levied under the Health and Safety at Work Act (HSWA) 1974 was in 1988, when BP was fined a total of £750,000 following three fatal accidents at the Grangemouth Refinery. The compensation paid by BP's insurers to the widows probably exceeded that amount. Such 'white-collar' crimes do not carry the stigma of, for instance, theft or criminal assault, probably because they are not often deliberate. Cynically, the Zeebrugge disaster was a commercial tragedy for Townsend-Thoresen because customers as well as employees were killed. Where the tragedy strikes the workers alone, the commercial damage is far less acute.

A series of disasters has made the public more aware of potential risks. The *Herald of Free Enterprise* capsize at Zeebrugge, the King's Cross Underground fire, the Clapham Junction railway accident, the *Piper Alpha* explosion in the North Sea and the many deaths at the Hillsborough football ground have led to public demand that enterprises be punished more effectively for endangering human life in the cause of profit. However, at the same time, industry has whole-heartedly espoused the philosophy of performance-related pay and individual appraisal of managers. There has been a move away from national collective bargaining in favour of decentralisation. A manager under constant pressure to cut costs and reduce the wages bill will be likely to employ more temporary and part-time workers, who are less well-trained and more at risk. He will also be encouraged to bring in subcontractors who will be unfamiliar with the day-to-day operations of his business. The Report of the Chief Inspector of Factories in 1985 showed that several companies had reduced the numbers of their in-house safety personnel.

> 'The current trends towards decentralisation within large corporations seems then to carry a real risk that decisions regarding health and safety issues will increasingly be the subject of narrow, short-term, cost-benefit considerations'. (James, 1992)

Since most would agree that the main aim of the law should be to prevent accidents, not to punish offenders, how can this best be achieved? The Robens Committee on Safety and Health at Work Report (1972), which expounded the philosophy that health and safety was a matter of common interest between management and workers, laid stress on the need to involve the workers in their own safety:

> 'We believe that if workpeople are to accept their full share of responsibility ... they must be able to participate fully in the making and monitoring of arrangements for safety and health at their place of work.'

To enable them to fulfil this role, they must have access to information. Also, in Sweden and the Australian State of Victoria, but not the UK, worker representatives have the power to order the employer to interrupt production while the situation is investigated by Government inspectors. In fact, there has been a recent decline in the militancy of British workers, often because of the fear of unemployment. The Chief Inspector of Factories gave an example of this:

'The extent of the fear of the loss of one's job may perhaps be gauged by the parents who offered to pay for damage to a conveyor when their 17-year-old son jammed his shovel into it to stop it after his arm became trapped at the tail drum. They offered to waive their claim for compensation if only their son could continue to work.' (*HSE Annual Report for the Manufacturing and Service Industries* (1983))

The enforcement agencies, the Health and Safety Executive (HSE), the local authorities and the trade unions complain that there are not enough inspectors. At a time of rising health and safety crime, numbers of 'policemen' have been reduced. Occupational accident statistics for 1981–85 show a marked increase in the total of reported major injuries for manufacturing and construction (Godard (1988)). There were over 700 deaths from accidents at work in 1985 and almost 20,000 major accidents. The number of fatal accidents to building workers in London was in 1987 the highest for more than twenty years. In addition, it has been suggested that up to 30 per cent of disease is work-related (Discher, Kleinman and Foster (1975)). An analysis by the Institution of Professionals, Managers and Specialists (IPMS), the union which represents HSE inspectors, disclosed in 1992 that there was about one inspector for every 1000 workplaces subject to inspection by the Factory Inspectorate division of the HSE. It also revealed that half of the premises had not been inspected for three years and that 69,564 had not been visited for eleven years (*The Third Alternative Report on the Work of the Health and Safety Executive* (IPMS, 1992)).

The most recent statistics (1991–2) show that employee fatalities have levelled off at around 320 a year and that the reported major injury rate for employees has been stable over the last three years. But this also reflects the decline of heavy industry in favour of the less hazardous service industries (since 1986 employment in coal mining has halved). The current fatal injury rate for employees in the construction sector is the highest of the employment sectors, five times higher than in manufacturing. The benign influence of legal regulation can be demonstrated by the fall in the number of head injuries on construction sites (by at least a quarter) after the coming into force in April 1990 of the Construction (Head Protection) Regulations 1989.

In the United States, the National Institute for Occupational Safety and Health (NIOSH) in 1983 developed a list of the ten main work-related diseases and injuries. The 'top ten' are as follows: occupational lung diseases, musculoskeletal injuries, occupational cancers, fractures, lacerations, eye injuries etc., cardiovascular diseases, disorders of reproduction, neurotoxic disorders, noise-induced hearing loss, dermatologic conditions and psychological disorders.

The Annual Reports of the Health and Safety Commission 1990–1 and 1991–2 contain recent statistics of occupational ill-health. The HSC points out the potential elasticity of the term, encompassing as it does conditions which are unequivocally work-related (lead poisoning, asbestosis) and conditions with multiple causes, some of which are occupational in origin (lung cancer in asbestos workers, 'sick building' syndrome). The latter can only be linked to the occupation by statistical means, demonstrating that the prevalence of the condition is higher among those exposed to the occupational factor. The most commonly reported diseases caused by work in Great Britain are general musculoskeletal conditions, hearing loss and stress/depression, and lower respiratory disease. All these conditions, as well as circulatory disease, are exacerbated by work in the case of thousands of workers.

If occupational health provision is poor and GPs are untrained in the diagnosis of work-related illness, many cases of occupational disease will never be identified. In practice, it is often difficult to draw a line between occupational and non-occupational disease. Diseases with long latent periods may escape detection if the worker has had several jobs over a number of years. The proliferation of new chemicals and combinations of chemicals may even now be causing harm. The International Labour Office estimates that it will take eighty years to assess the toxic properties of about 40,000 new chemical substances using current toxicological research methods.

The importance of statistics in this field makes it imperative that there be accurate reporting. The 1990 Labour Force Survey indicates that only about 30 per cent of non-fatal injuries are being reported to HSE. The Reporting of Injuries, Diseases and Dangerous Occurrences Regulations (RIDDOR) 1985 are proving difficult for employers to interpret (Chapter 3) and it is likely that proposals to strengthen the Regulations will be introduced. Another source of data is the figures for compensated industrial disease (Chapter 6), which show a rise in the number of claims for asbestosis, occupational asthma, mesothelioma, occupational deafness and vibration white finger. It must be emphasised that changes in compensation rules and benefit rates must be allowed for when examining trends.

Since the beginning of 1989 the Epidemiological Research Unit at the London Chest Hospital, in collaboration with the British Thoracic Society and the Society of Occupational Medicine and funded by the HSE, has operated a reporting system for cases of occupationally related respiratory disease seen for the first time by occupational and chest physicians throughout the UK. It bears the acronym SWORD (Surveillance of Work-related and Occupational Respiratory Disease). The success of the scheme has led to the introduction of a similar system for reporting occupational skin disease.

Representatives of the Commission of the European Communities are in the process of trying to secure agreement to a core of comparative statistics which will make it easier to make comparisons between member states. The HSE study (*Workplace Health and Safety in Europe* (1991)) suggests that British rates of fatal and non-fatal injury compare favourably with those of other European countries, but the unreliability of the British statistics makes this less compelling than it would otherwise be.

Evidence from all sources shows that it is in the small organisations that the worker is most at risk. The rate of major injuries is 50 per cent higher in firms employing fewer than 100 people than in large establishments. The rapid rise in the numbers of small businesses has exacerbated the problems of the inspectorate. Although it is important that new enterprises are not strangled with red tape soon after birth, it is obviously necessary to ensure that the law is obeyed throughout industry. To give inspectors more time to visit small businesses, the HSE in 1983 agreed an experimental policy of self-regulation with some of the large organisations in the construction industry. The Health and Safety Commission (HSC) proposed in 1985 (*Plan of Work for 1985–86 and Onwards*) that the HSE should withdraw from day-to-day inspection of common hazards and concentrate on areas of high risk. Self-regulation and control by private insurance companies should take its place. This move towards self-regulation reflected the policy of the Robens Committee, which was that the function of health and safety law is to provide a regulatory framework within which those in industry can themselves undertake responsibility for safety at work:

'The primary responsibility for doing something about the present levels of occupational accidents and disease lies with those who create the risk and those who work with them ... Our present system encourages rather too much reliance on state regulation and rather too little on responsibility and voluntary self-generating effect ... There is a role for government action. But those roles should be predominantly concerned with influencing attitudes and creating a framework for better health and safety organisation and action by industry itself.'

The philosophy of the Conservative Government of the 1980s and 1990s is to place more and more emphasis on self-regulation and deregulation. It conflicts with the dominant school of thought in the European Union which regards legislation as the most effective method of protecting the workforce. As a result, the call from the UK Government to abandon many technical and outmoded laws confronts the need to enact European directives, more and more numerous as the years go on. In 1993 the Government invited the Health and Safety Commission to review by 1994 all the legislation for which it is responsible. The review, advised by seven industrial Task Groups, was asked to examine whether there are burdens arising for business which are not offset by benefits and which could be eased without endangering necessary health and safety standards. In addition, there is movement in the European Union towards a more consistent approach to the enforcement of EC legislation by Member States. The EC Senior Labour Inspectors' Committee is discussing how common principles of inspection could be developed across the Community.

The philosophy of self-regulation was adopted in the recommendations of the public inquiry into the *Piper Alpha* disaster (*Cullen Report* (1990)). Operators of offshore installations should be required to submit to the HSE a safety case in respect of each of their installations. This should set out the safety objectives of the organisation, the safety management system, performance standards to be met and the means by which performance is to be monitored. The safety cases should be regularly updated and there should be periodic audits. The recommendations were enacted in the Offshore Installations (Safety Case) Regulations 1992. Failure to conform with the procedures specified in a safety case will be a criminal offence, unless the owner or operator can show that in the circumstances it was not in the best interests of the health and safety of persons to follow those procedures, or that the commission of the offence was due to a contravention by another company of its duty under the regulations to co-operate, and that the accused had taken all reasonable steps and exercised all due diligence to ensure that the procedures were followed. It is planned to extend the safety case idea to the railways when they are privatised.

The most extensive research so far undertaken on the operation of safety legislation, S. Dawson *et al.: Safety at Work: the Limits of Self-*

regulation: (1988), suggests that self-regulation unsupported by external enforcement is not as effective as the Robens Committee once supposed. Although in the long term it is in the interest of the employer to care for his workers, managers tend to be short-sighted when it comes to devoting resources to health and safety, especially in a period of economic recession.

Should more attention be paid to the responsibility of the individual manager and worker? The Chernobyl disaster led to prison sentences for the managers in charge. Up to now, prison sentences for breaches of the Health and Safety at Work Act are few in number and have all been suspended. Prosecutions for breaches of the health and safety laws can only be brought by the enforcement agencies. Inspectors find themselves in an invidious position. On the one hand, they desire to build a good relationship with employers; on the other, they are feared because of their powers of enforcement. An employer may be deterred from asking for the HSE's advice, realising that he may be forced to introduce expensive safety measures. The Health and Safety Commission, which is entrusted by Parliament with the overall policy aspects of health and safety at work, advises that prosecutions, expensive and time-consuming as they are, should be regarded as a measure of last resort. Research has shown that approximately one per cent of accident investigations by the factory inspectorate lead to prosecutions and, as might be expected, are more likely to follow from the investigation of an accident than from a routine inspection visit (Hutter and Lloyd-Bostock (1990)). Most prosecutions are brought in the magistrates courts, where legal costs are lower, but the sentencing powers of magistrates (up to £20,000 fine or imprisonment for six months) are paltry compared with the Crown Court. The chosen defendant in nearly all cases is the employing organisation; the number of prosecutions brought against individuals is small.

One sanction available against an individual is disqualification of a director of a company for up to two years for a breach of the health and safety laws (Company Directors' Disqualification Act 1986). The first such disqualification was imposed in 1992. Companies can purchase insurance for their directors to pay for the legal costs and expenses of defending a criminal prosecution, though not to pay any fines (Companies Act 1989, s. 137).

Where there has been a fatal accident, the Crown Prosecution Service (Procurator Fiscal in Scotland) may decide to prosecute for the common law crime of manslaughter. A private prosecution is also possible. Both the corporate body and the individual responsible for the death may be found guilty of this crime, the essence of which is gross carelessness, more culpable than run-of-the-mill negligence. Much will depend on the view the jury takes of the defendant's behaviour. The prosecution of the owners of the *Herald of Free Enterprise*, P. and O., and seven of its employees, in respect of the Zeebrugge disaster, failed because the judge ruled that there was not enough evidence against the company and its five senior personnel. The prosecution decided not to proceed against the other employees, the bosun and the captain. The company could have been convicted only if very senior employees, those who decided company policy, were guilty, because they were the 'directing mind' of the company.

One of the few examples of a conviction for manslaughter in the workplace occurred in Preston Crown Court in 1990 when a director of a small plastics company was convicted of the manslaughter of an employee who fell into an unguarded shredding machine. He was given a suspended prison sentence. The maximum penalty for manslaughter is imprisonment for life.

The Court of Appeal has given guidance on the degree of negligence which needs to be proved to a jury in order that it may return a verdict of manslaughter. In a medical case, an anaesthetist assisting at an eye operation failed to notice the disconnection for six minutes of the tube from the ventilator supplying oxygen. The patient suffered a cardiac arrest, from which he subsequently died. Expert witnesses described the conduct of the defendant as abysmal and a gross dereliction of care. The Court of Appeal held that he had been rightly convicted of manslaughter. Proof of the following states of mind in the defendant can properly lead the jury to make a finding of gross negligence:

(a) Indifference to an obvious risk of injury to health;
(b) Actual foresight of the risk coupled with the determination nevertheless to run it;
(c) An appreciation of the risk coupled with an intention to avoid it but with such a high degree of negligence in the attempted avoidance that the jury considers it justifies conviction;
(d) Inattention or failure to advert to a serious risk which goes beyond mere inadvertence in respect of an obvious and important matter which the defendant's duty demands he should address (as with the supply of oxygen to the patient). (*R. v. Adomako* (1993).)

In a second case heard at the same time an inexperienced houseman injected vincristine into the spine of a patient suffering from leukaemia, not realising that it should have been injected intravenously. Another junior doctor was supervising, but mistakenly believed that he was only supervising the lumbar puncture and did not need to check the drugs. The cytotoxic trolley was not in use, so the data chart was not available, and the senior nurse was not present. The patient died. The health authority, the doctors' employer, was clearly liable to pay compensation under the civil law, but it was held that the doctors were not guilty of manslaughter, because there were a number of excuses and mitigating circumstances which had to be taken into account. (*R. v. Prentice and Sullman* (1993).)

5.2   The Health and Safety Commission and the Health and Safety Executive

The Robens Committee in 1972 reported that the multiplicity of different agencies and inspectorates was an inefficient use of resources. They recommended a more unified and integrated system. The HSWA set up the HSC to undertake the functions *inter alia* of encouraging and organising research, making proposals for new legislation, setting and reviewing standards, approving Codes of Practice and generally laying down policy guidelines for the inspectorates. In the 1990s the growing pre-eminence of the European Community in matters of health and safety at work means that the Commission must be active in representing the United Kingdom's interests in EC negotiations and in advising on legislation to implement European directives.

The HSE is responsible to the HSC for the enforcement of the safety laws. After the *Piper Alpha* disaster in the North Sea in 1988 the HSE became the regulatory agency for the offshore oil and gas industry, taking over from the Department of Energy in April 1991. The Railway Inspectorate transferred to HSE from the Department of Transport in December 1990. The local authority environmental health officers police offices, shops and catering establishments. The forum for co-operation between HSE and local authorities is the HSE/Local Authority Enforcement Liaison Committee (HELA).

The Health and Safety at Work (N. Ireland) Order 1978 enacts similar provisions to the HSWA for Northern Ireland, where enforcement is through the Health and Safety Agency.

5.3　Health and safety statutes

The criminal law of health and safety is virtually all in the form of statute. It is not the purpose of this book to list all the detailed regulations. Encyclopaedic works like Redgrave, Fife and Machin: *Health and Safety* are the recognised source for this kind of information. The legislation falls into two broad categories:

(1) The 'old' legislation (pre-1974), including the Mines and Quarries Act 1954 (currently in the final stages of revision), the Agriculture (Safety, Health and Welfare Provisions) Act 1956, the Factories Act 1961 and the Offices, Shops and Railway Premises Act 1963.
(2) The Health and Safety at Work Act (HSWA) 1974, and statutory regulations made thereunder.

Eventually, all the old statutes will be reviewed and replaced where necessary with regulations under the HSWA, making a comprehensive code. The Provision and Use of Work Equipment Regulations 1992 and the Workplace (Health, Safety and Welfare) Regulations 1992 will eventually, when fully in force, replace much of the old legislation. The principal distinction between the old and the new legislation is that the former is confined to a particular kind of workplace, whereas the latter covers virtually all places of work. Work in a doctor's surgery falls outside legislation relating to factories, but the HSWA will apply.

The drafting of the old and new laws follows a different pattern. Traditionally, safety legislation spelled out legal requirements in some detail. Particular safety precautions were specified (scaffolding, crawling boards and so on) and temperatures and heights were laid down to the degree and the inch. The scope for lawyers to help their clients to avoid the legislation by interpretation was considerable.

> 'We seem always to have been incapable even of taking a general view of the subject we were legislating upon. Each successive statute aimed at remedying a single ascertained evil. It was in vain that objectors urged that other evils, no more defensible, existed in other trades or among other classes, or with persons of ages other than those to which the particular Bill applied' (Webb (1910)).

The HSWA in its central provisions is, in contrast, of apparent simplicity. Section 2 obliges the employer to ensure, *so far as is reasonably practicable*, the health, safety and welfare at work of all his employees. Section 3 places a duty on the employer to conduct his undertaking in such a way as to ensure, so far as is reasonably practicable, that people not employed by him who may be affected are not exposed to risks to their health or safety.

5.4 Reasonable practicability

The employer's duty under the general sections of the HSWA is to do that which is reasonably practicable. This is very similar to the duty to take reasonable care in the tort of negligence, discussed in Chapter 7. One difference is that section 40 HSWA provides that it is for the *defendant* to prove that it was not reasonably practicable to do more than he did, whereas the burden of proof in the civil law is on the plaintiff to show that the defendant was negligent. In a sense, the unspecific nature of the duty increases the power of the inspector who has more flexibility in identifying unacceptable hazards.

How can the employer discover what the law requires of him? Everyone has his own ideas of reasonableness, but even judges disagree. First, he can consult test cases to identify the factors which the courts will consider material. Cost benefit analysis plays an important part. In one case, an inspector ordered ASDA Stores to provide safety shoes free of charge for all its employees working in warehouses, as a precaution against having a foot crushed by a loaded roller truck. This would have cost £20,000 in the first year and £10,000 in each succeeding year. The company already provided safety footwear at cost price. There had been ten accidents in the previous year involving roller trucks in ASDA's 66 stores. The industrial tribunal disagreed with the inspector on the ground that the expense was disproportionate to the risk (*Associated Dairies* v. *Hartley* (1979)).

Are standards of reasonableness lower for small companies than for multinationals? A truly objective standard would apply the same criteria to all, whatever their financial position. There is, however, surprisingly little judicial guidance on this point. The following definition, a paraphrase of a dictum of Lord Justice Asquith in *Edwards* v. *NCB* (1949), appears in Redgrave:

'"Reasonably practicable" is a narrower term than "physically possible" and implies that a computation must be made in which the quantum of risk is placed in one scale and the sacrifice, whether in money, time or trouble, involved in the measures necessary to avert the risk is placed on the other; and that, if it be shown that there is a gross disproportion between them, the risk being insignificant in relation to the sacrifice, the person on whom the duty is laid discharges the burden of proving that compliance was not reasonably practicable. This computation falls to be made at a point of time anterior to the happening of the incident complained of.'

This fails to address the issue of whether the small employer can plead his poverty as a material factor in measuring the degree of sacrifice. My view is that lack of resources should not be a defence, because employees of small businesses would then have even less protection than they enjoy at present.

In some important areas, the employer is given detailed guidance as to what is reasonable in the form of an Approved Code of Practice. The Codes are drawn up after extensive consultation with the Government and both sides of industry and must be approved by the HSC. Breach of an Approved Code is not in itself a crime, but the failure on the part of any person to observe the Code is *prima facie* evidence of guilt, so that in practice it will be difficult to avoid a conviction if the recommendations in the Code have not been followed. Only Codes officially approved by the HSC have this special status. Additional guidance provided by the HSE does not count as an Approved Code, but is likely to be treated as persuasive by courts of law.

The duty under health and safety statutes is not always to do that which is 'reasonably practicable'. A more onerous duty is to do that which is 'practicable'. It is practicable to take a precaution which is possible in the light of current knowledge, even though the cost far outweighs the risk. Occasionally, the obligation is 'strict': liability is imposed for conduct which is neither negligent, nor deliberate, nor reckless (see Chapter 7).

English lawyers have become familiar with the concept of reasonable practicability which is closely allied with the common law idea of the reasonable man. Legislation emanating from the European Community employs different expressions. It remains to be seen how English courts will interpret phrases like 'Every employer shall make a suitable and sufficient assessment of the risks to the health and safety of his employees to which they are exposed while they are at work' (Reg. 3, Management of Health and Safety at Work Regulations 1992). It seems likely that they will find an employer guilty of an offence only where he has acted unreasonably in the measures he has taken or failed to take.

5.5   *Civil liability*

The action for breach of statutory duty is discussed at length in Chapter 7. It is necessary, however, to note here that the courts have in general construed the pre-HSWA legislation as giving a right to sue for damages in the civil courts to any person injured by breach of the statute. The obligations in sections 2–8 HSWA do not give rise to such a civil action (section 47). However, breach of a duty imposed by health and safety *regulations* is actionable in the civil law except in so far as the regulations otherwise provide. A civil action may be brought, for example, for a breach of the Asbestos Regulations 1987 or the COSHH Regulations 1988, but not for the Management of Health and Safety at Work Regulations 1992.

Someone injured by an employer's failure to take reasonable precautions may be involved in both criminal and civil proceedings. More frequently, the inspectorate decides against a criminal prosecution, leaving the victim to pursue an action in the civil courts, as in the case of the Abbeystead pumping station explosion in 1984. In the rare case where a criminal conviction is secured, the victim will be able to rely on the findings of the criminal court as proof of negligence. No sensible insurance company would fight a claim for compensation by a worker whose employer had just been convicted of failing to do that which was reasonably practicable. Even an investigation by the inspectorate into the causes of an accident is advantageous to victims, because public funds will pay for the costs of procuring technical evidence, on which individuals may rely, assuming that it is made available to them. Section 28(9) HSWA permits an enforcing authority at its discretion to furnish to any person who appears to be likely to be a party to any civil proceedings arising out of any accident or occurrence a written statement of relevant facts observed in the course of exercising statutory powers. If the discretion is not exercised, a plaintiff could apply to the courts for an order to disclose, as described in Chapter 3. Only if the enforcing authority could show that it would not be in the public interest to reveal information (e.g. because it concerned technical details about a nuclear submarine) would it be able to resist a court order.

The Criminal Injuries Compensation Board is empowered to pay damages out of public funds to the victims of crime. It is now governed by the Criminal Justice Act 1988. The scheme is not normally relevant in cases of work-related disease or injury, because it is confined to crimes where it is necessary for the prosecution to prove that the defendant either intended to cause personal injury to another or was reckless as to whether death or personal injury were caused. None of the crimes established by health and safety statutes come within this category. If, however, injury has been suffered or death caused because of someone's intentional act or gross negligence, the Crown Prosecution Service or a private individual might be able to prosecute for an offence under the Offences against the Person Act 1861, or for manslaughter, in which event the victim or his dependants would have a claim against the fund. Also, workers like nurses, social workers or security staff who have been attacked in the course of their work would be entitled to compensation from the fund. The Criminal Injuries Compensation Board can make a payment whether or not a criminal prosecution has in fact been brought. Assaults committed by those who cannot be convicted of a crime because their mental state or age precludes it nonetheless fall within the scheme.

The not insubstantial numbers of train drivers who suffer nervous shock when suicidal persons throw themselves on to the track in front of the train have now been given the right to compensation. The Act extends to UK territorial waters, British ships and aircraft, and to oil rigs etc. within the Continental Shelf Act 1964.

5.6 The powers of the inspectorate

The HSWA substantially increased the legal powers of the inspectorate in respect of the enforcement of all the safety legislation, old and new. Section 20 gives the power:

(1) to enter premises without the permission of the occupier (if necessary, accompanied by a policeman);
(2) to make such examination and investigation as may be necessary;
(3) to direct that premises or any part of them be left undisturbed for as long as is reasonably necessary for the purpose of any examination or investigation;
(4) to take measurements, photographs and recordings;
(5) to take samples of articles, substances and of the atmosphere and to take them away for examination;
(6) to require any person whom he has reasonable cause to believe to be able to give any relevant information to answer such questions as the inspector thinks fit to ask, and to sign a declaration of the truth of his answers (though this will not be

admissible in any proceedings against the person making the statement);

(7) to inspect and take copies of any books or documents which by virtue of any relevant statutory provision are required to be kept and any other books or documents which it is necessary for him to see for the purposes of any examination or investigation;

(8) to require any person to afford him such facilities and assistance as are necessary to enable him to exercise his powers;

(9) any other power which is necessary.

There is no exemption in this section for health professionals or confidential medical records. Only lawyers' confidential files are exempt from inspection and seizure.

Where the inspector in the course of his inspection finds what appears to him to be a contravention of a safety statute, or comes upon a state of affairs which he considers highly dangerous, he may issue an improvement or a prohibition notice. These powers, conferred by the HSWA, have proved popular with inspectors, if not with employers. An improvement notice must specify an alleged contravention of a relevant statutory provision. It orders the recipient to take remedial steps, as to guard a machine or provide improved seating at a supermarket checkout, within a specified period. A prohibition notice orders the discontinuance of an activity if the inspector states that, in his opinion, it does or will involve a risk of serious personal injury, whether or not accompanied by a contravention of a relevant statutory provision. In cases of imminent danger, the notice may be of immediate effect. Thus, the inspector has power to halt production in a bad case. In 1991/2 12,382 notices were issued, 8,377 improvement and 4,005 prohibition, slightly fewer than in 1990/1, but comparing with 9,480 in 1986/7.

The employer who feels that the inspector is being high-handed and demanding excessive precautions may appeal against either notice to an industrial tribunal, which has power to cancel, affirm or modify the notice. Pending the hearing of the appeal, all except immediate prohibition notices are suspended. Unlike other proceedings before industrial tribunals, costs are normally awarded against the loser to the party who wins the case. Tribunals are reluctant to overturn improvement notices issued for failure to implement specific statutory requirements. In *Harrison (Newcastle-under-Lyme)* v. *Ramsey* (1976), a notice that an employer should comply with his obligation under the Factories Act to paint his walls was upheld, despite the absence of any danger to health and the company's financial difficulties. There is more scope for argument when the employer is obliged to do that which is reasonably practicable and there is no Approved Code of Practice. In *West Bromwich Building Society* v. *Townsend* (1983), an environmental health officer served an improvement notice alleging that a building society was in breach of its duty to do what was reasonably practicable to protect its employees against robbers. It was ordered to fit anti-bandit screens. The tribunal in confirming the notice disregarded evidence that there was a difference of professional opinion about the value of such screens. On appeal to the High Court, the improvement notice was quashed.

The employer who fails to exercise his right of appeal to the tribunal may be prosecuted for failing to comply with the notice. At this stage, it is too late to argue that the issue of the notice was unjustified.

Prosecutions for offences under the health and safety statutes can only be brought by an inspector or by or with the consent of the Director of Public Prosecutions (in Scotland, prosecutions are brought by the Procurator Fiscal). Trade unions or individual employees or their families cannot bring private prosecutions. Prosecutions are reserved by the inspectorate for offences of a flagrant, wilful or reckless nature. In 1991/2 2,407 prosecutions were brought of which 2,122 secured a conviction. The average penalty per conviction was a fine of £1,134 (including fines of £250,000 against British Rail and £100,000 against both Shell and British Gas). Some offences under the Acts are triable only in the magistrates' courts, but most are triable either way, i.e. before the Crown Court or the magistrates. If the inspector chooses to prosecute in the Crown Court, this is because the offence is regarded as particularly serious. The Crown Court has power to order the payment of an unlimited fine and, in the case of a few offences like contravention of a prohibition or improvement notice, to send the offender to prison for up to two years. The magistrates can fine up to £20,000 for breach of s. 2–6 HSWA, breach of an improvement or prohibition notice or court remedy order, and imprison for up to six months for breach of an improvement or prohibition notice or court remedy order (Offshore Safety Act 1992). In respect of other offences the maximum fine available to the magistrates is £5,000.

Most prosecutions are brought against a corporate body. Because the corporation has no physical existence, it can only be convicted of a crime if a director or manager acting on its behalf has transgressed. Section 37 HSWA provides that where an offence has been committed by a body corporate 'with the consent or connivance of, or to have been attributable to any neglect on the part of, any director, manager, secretary or other similar officer of the body corporate', both the corporation and the manager can be guilty of an offence. The trade unions believe that too few prosecutions are brought against managers. The fear of imprisonment is more powerful a deterrent than a fine levied on the company.

The wording of s. 37 is repeated in s. 23 of the the Fire Precautions Act 1971 which was considered in *R* v. *Boal* (1992). Francis Boal was employed by Foyles as assistant general manager of its bookshop, but had been given no training in management or health and safety at work. On a day when he was in charge of the shop serious breaches of the Fire Precautions Act were discovered. Both he and his employer were prosecuted in the Crown Court. The Court of Appeal quashed his conviction (he had pleaded guilty on bad legal advice). These Acts intend to fix with criminal liability only those who are in a position of real authority, who have both the power and the responsibility to decide corporate policy, not those who have charge of the day to day running of the business. Boal was not a manager in the sense intended by Parliament. The conviction of Foyles was upheld.

5.7 Crown immunity

There is a common law rule that Acts of Parliament do not bind the Crown (the central Government and bodies acting on its behalf) unless the Crown is included either by an express provision or by necessary implication. Because Parliament is sovereign, it always has the power to extend the effects of legislation to the Crown, but many criminal statutes have not been so extended. This has allowed the employers of many thousands of workers, like the civil service and the health authorities, to escape the full effects of legislation. In the civil law, the ancient common law rules relating to the immunity of the Crown from actions for compensation were abolished by the Crown Proceedings Act 1947. This had the effect that in general an individual injured by the negligence of a Crown servant could claim compensation from the Crown in the civil courts, with a few exceptions, principally actions by members of the armed forces. The latter exception has now been removed for cases of injury occurring after the Crown Proceedings (Armed Forces) Act 1987 came into force.

Local government authorities and the boards of nationalised industries have no such special status and can be held liable in criminal or civil proceedings in the same way as a company in the private sector.

In the field of statutes relating to health and safety at work, the Factories Act 1961 but not all the provisions of the HSWA 1974 applied to the Crown. The duties imposed by the HSWA, and by regulations made thereunder, bind the Crown, but not the sections giving power to inspectors to inspect, issue improvement or prohibition notices, and prosecute. The exclusion from the enforcement provisions of large employers like health authorities and Government Ministries caused much criticism. A system of Crown Enforcement Notices was introduced as a compromise in 1978. HSE inspectors could serve these on Crown bodies in circumstances where a statutory improvement or prohibition notice would have been served, but they were in reality only advisory. Health authorities lost their Crown immunity in the National Health Service and Community Care Act 1990. Other Crown bodies, however, still enjoy Crown immunity, though individual managers and employees, like occupational health personnel employed in Crown establishments, can be prosecuted for their own breaches of duty. Thus, the Home Office could not be prosecuted for an unsafe system of work in one of HM Prisons, but the prison doctor could be prosecuted if he had personally failed to take such precautions as were reasonably practicable to protect the staff or prisoners from risk to their health.

5.8 General duties under the Health and Safety at Work Act 1974

The old legislation placed duties for the most part on the occupier of premises. The HSWA spreads its net more widely. The following is a brief summary of the general duties imposed by the Act.

(1) *Section 2* makes it the duty of every employer to ensure so far as is reasonably practicable, the health safety and welfare at work of his employees. 'Employees' includes persons who work under a contract of employment or apprenticeship. The Health and Safety (Training for Employment) Regulations 1988 provide that trainees are to be treated as employees at work, except those receiving training at an educational establishment. Students in schools and colleges will have to rely on the provisions of the Act which require the employer to take precautions to protect those not employed by him who may be affected by his activities. Homeworkers will only be covered by s. 2 if they are classed as employees.

The employer is obliged to provide and maintain so far as is reasonably practicable safe plant and equipment. This extends to work done away from the employer's premises, if the employer could be reasonably expected to exercise control over it. Any place of work under the employer's control must be maintained in a reasonably safe condition. The employer must also maintain reasonably safe systems of work. This may impose an obligation on an employer in charge of a site on which a number of different contractors are working to co-ordinate their activities. This duty is now made express in the Management of Health and Safety Regulations 1992, Reg. 10.

In *R* v. *Swan Hunter* (1981), a fire killed eight men working on a ship in the Swan Hunter shipyard. The fire was caused by a welder who, without any negligence on his part, ignited leaking oxygen with a welding torch. An employee of a subcontractor, Telemeter Installations, had failed to turn off the oxygen supply in a confined space when he left work on the previous evening. Swan Hunter had taken reasonable care to inform and train their own employees as to the dangers of oxygen, but they had not concerned themselves with the instruction or training of those not employed by them but working on their site. The Court of Appeal upheld the convictions of both Swan Hunter and Telemeter for breaches of s. 2 HSWA:

'If the provision of a safe system of work for the benefit of his own employees involves information and instruction as to potential dangers being given to persons other than the employer's own employees, then the employer is under a duty to provide such information and instruction, so far as is reasonably practicable.'

The Construction (Design and Management) Regulations require employers or project supervisors to appoint health and safety co-ordinators at all construction sites and to prepare a construction plan. The principal contractor, as he is termed, will be responsible for co-ordinating the activities of all the contractors on the site and also the provision and use of shared work equipment. These measures do not absolve the individual contractor from responsibility.

The duty to provide safe plant does not depend on whether the plant is in use. In *Bolton MBC* v. *Malrod Installations* (1993) a contractor engaged to strip asbestos had installed a decontamination unit. The day before he started work an inspector found electrical faults. It was held that the defendant was under a duty to ensure that the decontamination unit would be safe when the employees came to use it and he was convicted of an offence under the HSWA.

The employer must also make arrangements for ensuring, so far as is reasonably practicable, safety and absence of risks to health in connection with the use, handling, storage and transport of articles and substances. He must provide such information, instruction, training and supervision as is necessary to ensure, so far as is reasonably practicable, the health and safety at work of his employees. He must maintain and provide a working environment for his employees that is, so far as is reasonably practicable, safe, without risks to health, and adequate as regards facilities and arrangements for their welfare at work. It is arguable that this could oblige the employer to provide medical or nursing services in a particular case. The civil courts have held that an employer who failed to institute regular examinations of his workers who had been exposed to carcinogenic substances failed in his common law duty to take reasonable care, despite the lack of any specific statutory obligation (*Wright* v. *Dunlop Rubber Co.* (1972), Chapter 4).

(2) *Section 3* obliges both employers and self-employed persons to do what is reasonably practicable to protect persons not employed by them who may be affected by their activities. Thus, a place of work and work activities should be made reasonably safe for employees of contractors, other visitors, and members of the public in the vicinity (the duty is not confined to those present on premises occupied by the employer). When a passer-by was killed by the collapse of a building in the course of demolition into the street, the company in control of the site was convicted of an offence under s. 3 HSWA, as was the BBC after a member of the public who had volunteered to take part in a stunt for a TV programme was killed when a safety rope failed.

In *R* v. *Mara* (1987), a small cleaning company agreed to clean the premises of International Stores in Solihull. John Mara was a director of the company, of which the only other director was his wife. His company provided a polisher/scrubber which was left on the premises for the use of Mara's cleaners. It was agreed that the employees of International Stores could also use this machine. An employee of International Stores was electrocuted while using the machine which was seriously defective. Mara was prosecuted and convicted in the Crown Court of an offence of consenting or conniving to a breach by his company of s. 3 HSWA. He was fined £200.

Prosecutions under section 3 have been brought where the defendant is accused of spreading the bacterium *legionella pneumophila* (LP) in such a way as to cause risk to the general public. The most recent was *R* v. *Board of Trustees of the Science Museum* (1993). An inspection disclosed that these bacteria existed in the water in the air cooling system at the Science Museum. The trustees of the museum were convicted in the Crown Court of a s. 3 offence, in that they failed to institute and maintain a regular regime of cleansing and disinfection, failed to maintain in operation an efficient chemical water treatment regime and failed to monitor the efficacy of the regime, so that members of the general public were exposed to risks to their health from exposure to the bacteria. They appealed, arguing that there was no proof that members of the public inhaled LP, or even that LP had escaped into the atmosphere to be inhaled. The appeal was dismissed. There was a risk to the public, whether or not it could be proved to have materialised. The primary purpose of the HSWA was preventive, in comparison with the civil law which provides compensation only where the plaintiff can prove damage (Chapter 7). The defendants were fined £500 and ordered to pay prosecution costs of £35,000.

(3) *Section 4* relates not to employers but to 'controllers' of premises who make available non-domestic premises as a place of work to those who are not employed by them. It is the duty of the person in control of such premises to take such measures as it is reasonable for a person in his position to take to ensure, so far as is reasonably practicable, that the premises and means of access and egress thereto are safe and without risks to health. A lift in a block of flats, which needed repair by electricians, has been held to fall within this section (*Westminster CC* v. *Select Managements* (1985)).

(4) *Section 6* has been strengthened by amendments in the Consumer Protection Act 1987. Duties are imposed upon any person who designs, manufactures, imports or supplies any article for use at work, or manufactures, imports or supplies a substance for use at work. This part of the 1987 Act is aimed at the protection of workers, not consumers, unlike Part I which deals with product liability, discussed in Chapter 7. 'Substance' includes a microorganism. The safety requirements concern the safety of goods and substances in use, and in the course of setting, cleaning, maintenance, dismantling and disposal (goods) and handling, processing, storing and transporting (substances), by a person at work.

The duty is to do that which is reasonably practicable, whether by information, advice or otherwise. Risks to health and safety

which could not have been reasonably foreseen are to be disregarded in deciding whether there is a breach of the statute.
Reasonable testing and examination should be undertaken.
Manufacturers of articles or substances must carry out or arrange
for the carrying out of research necessary to discover any risks and
to minimise or eliminate them. The manufacturer, importer,
designer and supplier have a duty to take such steps as are
necessary to warn of risks to health and safety, the results of
relevant tests, and of any conditions necessary to ensure that the
article or substance will be safe. They must also take such steps as
are reasonably practicable actively to communicate relevant
revisions of information, when it becomes known after the product
has been supplied that there is a serious risk to health and safety.

Section 6 allows the designer, manufacturer, importer or supplier
of an article to obtain a written undertaking from the person to
whom the article is supplied that the latter will take specified steps
to ensure its safety (as, for example, that he will only use a machine
after a guard has been fitted or after it has been repaired). This
undertaking relieves the supplier from the obligation to ensure that
the article is safe when it leaves his hands, as long as it was
reasonable to rely on it.

The Supply of Machinery (Safety) Regulations 1992 will oblige
manufacturers who supply work equipment covered by the
regulations to comply with European Community safety standards
from 1 January 1995. Until then they must comply with British
standards.

The Chemicals (Hazard Information and Packaging) Regulations
1993, known as CHIP, require suppliers of chemicals to identify the
hazards of the chemicals they supply, to give information to those
to whom the substances are supplied and to package the chemicals
safely. There is a positive duty on suppliers to assess and classify.
The duty is to exercise due diligence. It is not sufficient to take the
manufacturer's classification on trust, but if you deal with him
regularly and use your experience and commonsense simple
checks in a reference book will be enough. Safety data sheets will
need to be provided for dangerous chemicals which are to be used
in connection with work. There is an Approved Code of Practice:
*Safety data sheets for substances and preparations dangerous for
supply.*

(5) *Section 7* states that it is the duty of every employee while at
work:

(a)   to take reasonable care for the health and safety of himself and of other persons who may be affected by his acts or omissions at work; and

(b) as regards any duty or requirement imposed on his employer or any other person by relevant statutory provisions, to co-operate with him so far as is necessary to enable that duty or requirement to be performed or complied with.

Note that this section imposes a duty on the employee to look after himself as well as his fellow employees. Employees may be prosecuted as well as or instead of their employer. Prosecutions are few. More often, an employer uses the sanction of dismissal for gross misconduct in a case of deliberate flouting of safety rules. Assuming he has followed a fair procedure, this is likely to be upheld as a fair dismissal by an industrial tribunal (see Chapter 8).

(6) *Section 8* makes it a criminal offence for any person intentionally or recklessly to interfere with safety arrangements. This is something more than mere carelessness: the prosecution must prove that the defendant showed a wanton disregard for the safety of others.

(7) *Section 9* prohibits any employer from charging his employees for any safety equipment provided in respect of any *specific* statutory requirement. Now that the Personal Protective Equipment Regulations 1992 oblige employers to provide PPE wherever necessary, the employer will never be permitted to levy a charge.

5.9 Safety representatives and safety committees

Although in the early part of the nineteenth century workers tacitly accepted poor conditions, concentrating their efforts on the improvement of pay and working hours, by 1871 the improvement of working conditions had become a definite part of trade union policy. The earliest safety statutes were, therefore, the product of what Sydney and Beatrice Webb described as 'middle class sentiment'. Trade unions eventually supported the need for legislation to lay down minimum standards, backed by a powerful inspectorate. Workers with weak bargaining power would be protected, and there would be no opportunity for employers to entice workers to work in unsafe conditions for extra pay.

Participation by employees in the policing of safety legislation has a long history in the mining industry. Since 1872, coalminers have had a statutory right to appoint representatives to inspect

mines on their behalf. The voluntary establishment in manufacturing industry of joint safety committees, with representatives
of management and workers, grew during the 1960s. The influence
of Heinrich, who in 1931 first postulated the thesis that accident
prevention should be based on prediction, rather than 'shutting the
stable door after the horse has bolted', by the 1970s had led to the
development of more systematic approaches to accident prevention by some large organisations, involving the setting of positive
safety policies and objectives in consultation with the workers. The
Robens Committee in 1972 wrote:

'There is no legitimate scope for "bargaining" on health and
safety issues, but much scope for constructive discussion, joint
inspection and participation in working out solutions.'

The Committee recommended that worker participation in
health and safety be put on a statutory basis so that all, not just
enlightened, employers would be compelled to involve the workforce. A fundamental political debate then arose. Should the new
health and safety legislation provide for management to consult
with trade unions representing workers, or with employee representatives elected by all the workers, not just those who were
members of the union? The Health and Safety at Work Act finally
became law after the election of a Labour Government. The Act
(s. 2(4)) provides that regulations may provide for the appointment
by recognised trade unions of safety representatives from among
the employees, and that those representatives shall represent the
employees in consultation with the employers. The Safety Representatives and Safety Committees Regulations and Approved
Code of Practice 1977 finally came into force in 1978. Employers
who do not recognise trade unions (and there is now no statutory
procedure by which recognition can be forced on the employer)
are, therefore, permitted to avoid the mandatory involvement of
safety representatives.

The safety representatives having been appointed by the recognised union, their names must be notified to the employer. They
should normally be employees with at least two years' service.
Numbers will depend on the nature of the workplace and the levels
of risk. It is the duty of the employer to consult them on matters of
health and safety. They have a right to reasonable time off with pay
for training (see Chapter 8).

As well as consultation with the employer, the safety representatives have the power to investigate potential hazards and
dangerous occurrences at the workplace. They have the right

routinely to inspect the workplace or part of it on giving reasonable notice in writing to the employer of their intention to do so, usually every three months, but more often if there have been substantial changes in working conditions, or new information from the HSC/HSE. In the event of a notifiable accident or disease, they have the right to inspect the scene as long as it is safe to do so.

The statutory regulations provide that a safety representative, although liable to prosecution as an employee under sections 7 and 8 HSWA, shall not be under any additional legal duty by virtue of exercising his functions as a safety representative. A representative who forgot to mention to the manager dangerous fumes reported to him by an employee would not be legally liable for that failure. A safety officer would have no such immunity.

Every employer, if requested in writing to do so by at least two safety representatives, must establish a safety committee. Safety committees should be concerned with all relevant aspects of health, safety and welfare of persons at work in relation to the working environment. An HSE Guidance Note advises that the functions of a safety committee could include:

(1) The study of accidents, diseases, statistics and trends, so that reports can be made to management with recommendations for improvement.
(2) Consideration of reports from the inspectorates and liaison with them.
(3) Consideration of reports from safety representatives.
(4) Assistance in the development of works safety rules and safe systems of work.
(5) A watch on the effectiveness of safety training.
(6) A watch on the adequacy of safety and health communications in the workplace.

The safety committee should be small, with at least as many worker representatives as management representatives. Occupational health physicians, industrial hygienists and safety officers should be *ex officio* members.

The trade unions would like the powers of safety representatives to be increased. They advocate that the safety committee should appoint occupational health personnel and control occupational health and safety services. The safety representatives should be given the power to issue a prohibition notice where they believe there is an imminent risk of serious injury or harm.

The obligation to appoint and recognise worker representatives was extended to the offshore oil and gas industry by the Offshore

Installations (Safety Representatives and Safety Committees) Regulations 1989). Election of safety representatives is done by the entire workforce. The installation manager must establish a system of constituencies of about 40 workers, each constituency to elect one representative.

New general provisions were added to the 1977 regulations by a Schedule to the Management of Health and Safety at Work Regulations 1992. Every employer must consult safety representatives in good time about the introduction of changes in the workplace, health and safety information, training and so on. He must also provide them with such facilities and assistance as they may reasonably require. The employer must not dismiss or discriminate against a safety representative who is doing his job in good faith. Statutory protection is conferred by the Offshore (Protection against Victimisation) Act 1992 and the Trade Union Reform and Employment Rights Act 1992 (see Chapter 8).

5.10   Access to information

There is no general principle of freedom of information enshrined in the laws of this country. On the whole, therefore, industry has no duty to give warning either to workers or to the general public of hazards created by the enterprise. There are at least three possible routes whereby the disclosure of such information may, exceptionally, be required:

(1)   the management of the enterprise may be placed under a statutory duty to notify a public authority which may then disseminate the information;
(2)   management or the inspectorates may be required by statute to notify workers directly or through their representatives;
(3)   a court may order disclosure of information relevant to legal proceedings.

Disclosure to a public authority is required by various legal regulations. The Reporting of Injuries, Diseases and Dangerous Occurrences Regulations 1985 (RIDDOR) have already been discussed in Chapter 3. The Notification of New Substances Regulations 1982, amended in 1991, and currently under review, oblige manufacturers or importers of new chemical substances to notify the HSE of technical details. The Dangerous Substances (Notification and Marking of Sites) Regulations 1990 require notification and marking of sites where there are 25 tonnes or more of dangerous substances present.

Membership of the European Community has brought with it obligations in connection with major hazards. After the Seveso disaster in Italy, a Directive was agreed which was incorporated into UK law as the Control of Industrial Major Accident Regulations 1984, amended in 1990. These apply to industrial activities involving very toxic, flammable or explosive substances, and to the isolated storage of quantities of such substances. Manufacturers who have control of such activities must identify major accident hazards and take adequate steps to prevent accidents, limit their consequences, and provide persons working on the site with the information, training and equipment necessary to ensure their safety. Where a major accident occurs, the manufacturer must notify the HSE, which in its turn must notify the European Commission. In addition, he must in any event give information to the HSE about the dangerous substance, a description of his process or storage arrangements, a map of the site, details of the workforce on the site, emergency procedures, and even prevailing meteorological conditions in the vicinity. The manufacturer must also approach the local authority with a view to informing persons outside the site of possible risks. These regulations are likely to be extended in 1994–95.

All these laws, however, direct that the information be not revealed to third parties without the consent of the organisation to which it relates. Section 28 HSWA forbids the disclosure of information obtained under s. 27 or furnished in pursuance of a statutory requirement. Certain exceptions are set out in s. 28, notably disclosure to a local authority, water authority or to the police. Disclosure to occupational health personnel is not permitted.

Inspectors exercising powers of inspection may not as a general rule disclose any information obtained by them other than for the purpose of legal proceedings, except that they are expressly authorised by s. 28 (8) to disclose information to employees or their representatives, as well as to the employer, if it is necessary for the purpose of keeping them adequately informed about matters affecting their health, safety or welfare. This may be either factual information obtained as to the premises or anything done therein, or information with respect to action taken or to be taken by the inspectorate. The Environment and Safety Information Act 1988 compels the inspectorates to make a list of improvement and prohibition notices available to the general public.

Section 2 HSWA imposes a duty on the employer to provide such information to his employees as is necessary to ensure, so far as is reasonably practicable, their health and safety at work. The Management of Health and Safety at Work Regulations 1992 oblige the employer to tell his employees of any special health risks involved in their work. Section 6 HSWA obliges manufacturers, importers, suppliers etc. of an article or substance to give adequate information about any conditions necessary to ensure that it will be safe and without risks to health when properly used. The Classification, Packaging and Labelling of Dangerous Substances Regulations 1984, amended 1989, obliged the supplier of certain specified substances to put warning labels on the containers, stating the classification of the particular hazard (e.g. toxic, corrosive, highly flammable), safety precautions, and the name of the substance and of its supplier or manufacturer. They are now replaced by the Chemicals (Hazard Information and Packaging) (CHIP) Regulations 1993 which regulate both suppliers and consignors by road of dangerous chemicals.

The Safety Representatives Regulations 1977 require the employer to make available to safety representatives information necessary to enable them to carry out their functions. There is a number of exceptions, including a document consisting of or relating to any health record of an identifiable individual. The Approved Code of Practice particularises the sorts of information which should be made available to safety representatives by the employer. It includes plans about proposed changes, technical information about plant, equipment and substances, information kept by the employer to comply with RIDDOR, and any other information specifically related to matters affecting the health and safety at work of his employees, including the result of any measurements taken by the employer or persons acting on his behalf in the course of checking the effectiveness of his health and safety arrangements.

In practice, if the employer and the inspectorate stubbornly refuse to disclose information, the only method of obtaining it is for an individual to bring a civil action for damages in the courts and apply for discovery of documents (see Chapter 3). Where no actual damage has been suffered by any individual, this option will not be available.

5.11 The control of substances hazardous to health

Much of the work of health professionals in British industry has from the nineteenth century onwards been concerned with the prevention of disease caused by exposure to dangerous substances.

Lead poisoning in the pottery trades, necrosis of the jaw caused by working with phosphorus in the match factories, mercury poisoning in the process of silvering mirrors and, of course, the effects of the inhalation of dust of various kinds, were well-known and documented long before the eventual creation of statutory controls. 'Phossy-jaw' was mentioned in Parliamentary Reports of 1843. Legislation prohibiting the use of white phosphorus was finally passed only in 1908.

The development of the chemical and pharmaceutical industries and the discovery of nuclear power in the twentieth century has greatly added to the potential risks faced by the workers. The horror of this kind of danger is that irreparable damage may be suffered before the risk is identified, because the symptoms may only appear years after exposure. Even today, legislation may be slow in gestation. The Health and Safety Commission is a tripartite body. It has to reconcile the interests of commercial industry and agriculture and the trade unions. Research has to be commissioned. In the 1990s the control of dangerous substances is seen as an international problem and laws increasingly originate from Brussels rather than London. Control of dangerous substances at work is linked with pollution control to protect the public and the environment.

Up to now, legislation has been passed on a substance by substance basis. The principal regulations which in practice will most concern OH professionals are as follows:

(1)   The Coal Mines (Respirable Dust) Regulations 1975 and 1978 (currently under review).
(2)   The Control of Lead at Work Regulations 1980 (and Approved Code of Practice 1985).
(3)   The Ionising Radiations Regulations and Approved Code of Practice 1985.
(4)   The Control of Asbestos at Work Regulations and Approved Code of Practice 1987.

Just as the HSWA introduced a new philosophy by providing general rules for all places of work, the Control of Substances Hazardous to Health (COSHH) Regulations and Approved Codes of Practice extend to all hazardous substances whether or not specifically named. First, they require the employer to assess the risk. Secondly, he must introduce engineering controls to reduce exposure to the lowest level which is reasonably practicable. Upper limits may be set by the regulations themselves. Where necessary, personal protective equipment must be provided as a 'back-up', but never as a replacement for proper engineering. Thirdly, the premises and perhaps also the workers must be monitored to check that controls are effective. Fourthly, where exposure may cause harm to health, health surveillance procedures, ranging from keeping a health record of exposure to regular checks by a doctor, must be instituted. Some regulations, like the Control of Lead at Work (1980), Ionising Radiations (1985) and Control of Asbestos at Work (1987) direct that employees who will be exposed to the dangerous substance shall be medically examined before or soon after the beginning of their employment in order to provide a base-line for later tests. This is not specifically mentioned in the COSHH Regulations, but may be advisable in some cases.

The 'one-substance' regulations have by definition already identified the hazardous substance. One of the revolutionary aspects of the COSHH Regulations is that it will to some extent be up to the employer to assess the danger. He will no longer be able to wait for an official Government warning. Regulation 6 provides that an employer shall not carry out any work which is liable to expose any employees to any substance hazardous to health unless he has made a 'suitable and sufficient' assessment of the risks created by that work to their health and of the steps needed to meet the requirements laid down in the Regulations. The employer is not left completely on his own. The Regulations list a number of sub-stances defined as hazardous to health. They have already been referred to in Chapter 4. They include any substance classified as being very toxic, toxic, harmful, corrosive or irritant under the Classification, Packaging and Labelling of Dangerous Substances Regulations 1984, and those substances which have a maximum exposure limit (MEL) (as specified in Schedule 1 of the COSHH Regulations), or occupational exposure standard (OES) approved by the HSC. (A MEL is the maximum limit of exposure considered safe which should never be exceeded, an OES is the acceptable limit within which the employer should aim to stay as a general rule.)

This is not, however, the end of the list, for employers will have to ask themselves whether their workers are exposed to any *other* substance which might be harmful. The HSC advises that the following questions should be answered:

(1) What hazardous substances are there?
     In most cases, the employer will not need to commission his own research, but will be able to rely on information from the supplier, including labels attached to the containers, trade and technical literature, HSE Guidance Notes and data sheets, and

past experience. In the case of research laboratories themselves, new and unknown substances will have to be treated with great caution until their properties become known.

(2) What are the harmful effects?
The effect will partly depend on the form of the substance and partly on the way it acts on the body. Answers will be obtained from the same sources as in (1) above. The possibility of dangers arising from a mixture of substances should be considered.

(3) Where will the hazardous substance be likely to be present?

(4) Who may be affected?

(5) How great is the degree and length of exposure likely to be?
The employer must enquire into working practices and experience. All routes must be considered – inhalation, and contamination of skin, clothing, food, drink and smoking materials. Exposure sampling may be necessary. This may be an examination of the working environment or of the employees themselves, or both. The Code of Practice states that '... it is essential that any sampling exercise should be planned, and the results considered, by someone with sufficient knowledge of occupational hygiene principles.'

(6) How do the exposure data compare with recognised standards?

(7) What action should be taken in response to the assessment?
Control is the most important message. Perhaps the substance can be eliminated from the process, or replaced. Efficient engineering controls are the next consideration. In all cases, control should be achieved by measures other than personal protective equipment, if reasonably practicable. To ensure that everything is going according to plan, routine monitoring of the environment, or health surveillance may be necessary. Assessments should be recorded and made available to the employees. They should be reviewed if there is a change in the work, or new information comes to light.

The trade unions argue that substances should no longer be regarded as 'innocent until proven guilty':

'The cost of giving inanimate substances the benefit of a criminal court's defence has been unacceptably high, with 2,000 asbestos deaths a year standing as a tragic testament to this traditional approach. The histories of mineral oil, silica, coal dusts, gas retort fumes and chemicals in the rubber and dyestuffs industries provide similar examples of the high human and community

costs of giving substances, not workers, the benefit of the doubt.'
(General, Municipal, Boilermakers and Allied Trades Union
(1987).)

The duties imposed on the employer by the Regulations are
owed in the first instance to his own employees and they are strict:
he *must comply* with the Regulations, not merely to do that which is
reasonably practicable, though the use of wording like 'suitable
and sufficient assessment' and 'suitable health surveillance' allows
inspectors and judges flexibility in interpretation. Also, Reg. 16
states that in any proceedings for an offence consisting of a contravention of the Regulations it shall be a defence for any person to
prove that he took all reasonable precautions and exercised all due
diligence to avoid the commission of the offence. In other words,
the burden of showing that reasonable care was taken is on the
employer, not the prosecutor. As the Regulations do not exclude a
civil action for breach of statutory duty, it is likely that they will
give rise to a number of claims for damages for breach of statutory
duty as well as criminal prosecutions. There is an important distinction between civil and criminal proceedings. Section 47 (3)
HSWA states that no defence made available by health and safety
regulations in respect of criminal proceedings shall afford a
defence in civil proceedings, unless specifically extended to civil
liability. The 'due diligence' defence in Reg. 16 does not apply to
civil actions for breach of statutory duty, therefore the civil liability
of the employer under several of the Regulations, at least to his
employees, is strict.

The employer also has obligations under the Regulations to non-
employees. The duties to assess risks (Reg. 6), prevent or control
exposure (Reg. 7) and use control measures, and maintain, examine
and test control measures (Regs 8 and 9) are owed both to non-
employees on the employer's premises (e.g. contractors, students)
and other persons likely to be affected by the employer's work (e.g.
members of the public living nearby). The employer is only obliged
to take care for these persons so far as is reasonably practicable (he
would still have the burden of proving that it was not reasonably
practicable to do more than was in fact done (s. 40 HSWA)).

The duty to monitor exposure at the workplace (Reg. 10) and to
provide information and training (Reg. 12) is owed to employees
and non-employees on the employer's premises, but not to persons
off the premises. Again, the duty to non-employees is only to do
that which is reasonably practicable. *The duty of health surveillance
(Reg. 11) is owed only to employees* (see Chapter 4).

Take a typical scenario of a factory manufacturing hazardous chemicals operated by Company X. Several different categories of people are affected by X's activities. On the factory premises, there are:

(1)   X's own employees;
(2)   employees of Company Y, which has a contract with X to instal new electrical wiring;
(3)   students from the nearby university who are visiting the factory;
(4)   firemen, who have been summoned to deal with a small fire caused by defective wiring.

As well, the factory is near several other factories and a motorway.

Company X's management must comply with all the regulations in respect of their own employees. Also, they will have to do that which is reasonably practicable to assess the risk and control exposure to hazardous substances of all those in the area, both on and off their premises, but there is no obligation to monitor any exposure outside the factory. Reasonable information needs to be given to the employees of Company Y and the students, as well as to the firefighters, but not to the public off the premises. A health record, and other health surveillance procedures, will be required only for those directly employed by X.

What are the obligations of Company Y to its electricians, who are its employees? Of course, it has general duties under the HSWA itself to give training and take reasonable precautions against foreseeable dangers. Reg. 2 (2) provides that any reference in the Regulations to an employee being exposed to a substance hazardous to health is a reference to the exposure of that employee to such a substance *arising out of or in connection with work which is under the control of his employer*. Whether Company Y has duties under the COSHH Regulations depends on whether the exposure of its employees to a hazardous substance arises out of or in connection with work which is under the control of Company Y. Is the 'work' referred to the manufacture of the chemicals, or is it the installation of the wiring? The Code of Practice, paragraph 5, states that 'where the employee of one employer works at another employer's premises both employers will have duties under the Regulations'. However, it then proceeds to indicate that 'it will usually be appropriate if the employer *having control* of the work undertakes exposure monitoring', which implies that the other employer is not in control. In my view, the Regulations must refer to the work which creates the hazard, so that Company Y would not have any obligations under COSHH. Similarly, if Company X is a travel agency, and Company Y does work on Company X's premises using a substance hazardous to health, the COSHH Regulations apply to Company Y, but not to Company X. Obviously, cooperation and collaboration between employers will be necessary. If this is correct, the duty to maintain a health record applies exclusively to workers who are both employees and exposed to a substance hazardous to health by work under the control of their own employer.

Reg. 3 states that the Regulations shall apply to self-employed persons, but not to the master or crew of a sea-going ship or the employer of such persons, in relation to normal ship-board activities. They do not cover lead, ionising radiations or asbestos, where regulations are already in force. They do not extend to work underground in a mine, because new respirable dust regulations are proposed under the Mines and Quarries Act. Nothing in the Regulations shall be taken to prejudice any requirement imposed by public health or environmental protection legislation. Although the Regulations cover exposure to micro-organisms, this is merely when the risks of exposure are work-related, not incidental (Reg. 2 (2), Chapter 4). If I am exposed to a dangerous substance in the course of having it administered to me as part of my medical or dental treatment, or as part of a research study, the Regulations do not apply, unless there is also a risk to the health professional caring for me.

Reg. 7 provides that employers must either prevent the exposure of their employees to hazardous substances, or, where not reasonably practicable, adequately control it. So far as is reasonably practicable, this should be secured without relying on protective equipment, but if this is not possible (e.g. routine maintenance, plant failure, lack of technical feasibility), the employer must provide suitable personal protection in addition to reasonably practicable engineering controls. Items of protective equipment should comply with the requirements of the International Standards Organisation, British or European standards, where they exist. Where there is exposure to a substance for which a MEL is specified in Schedule 1, the control of exposure is adequate, so far as the inhalation of that substance is concerned, only if the level of exposure is reduced at least to below the MEL, and even lower if reasonably practicable. Where exposure is to a substance for which an OES has been approved by the HSC, the control of exposure is adequate, so far as the inhalation of that substance is concerned, if either the OES is not exceeded, or where it is exceeded, the employer identifies the reasons and takes appropriate action to remedy the situation as soon as is reasonably practicable.

The MEL is the maximum concentration of an airborne substance, averaged over a reference period, to which employees may be exposed by inhalation under any circumstances. Control to the OES, or below it, can always be regarded as adequate, so far as exposure from inhalation is concerned. The absence of a substance from the lists of MELs and OESs does not indicate that it is safe. Some substances are hazardous by routes other than inhalation, and lists will always need updating. Where a substance has not been listed, the employer should control exposure to a level to which the bulk of the population could be regularly exposed without injury.

All kinds of control measures should be considered, ranging from totally enclosed process and handling systems to the provision of ventilation and adequate facilities for washing and laundering clothing. Control measures must be regularly examined, maintained and tested, and records must be kept. Reg. 8 obliges the employer to take all reasonable steps to ensure that control measures and personal protective equipment are properly used and applied. 'Every employee shall make full and proper use of any control measure, personal protective equipment or other thing or facility provided pursuant to the Regulations and, if he discovers any defect therein, he shall report it forthwith to his employer.' The employee who contravenes this Regulation is committing a crime. Also, his employer may decide to treat a bad case as a disciplinary matter.

Reg. 10 concerns the monitoring of exposure at the workplace. Monitoring means 'the use of valid and suitable occupational hygiene techniques to derive a quantitative estimate of the exposure of employees to substances hazardous to health'. It is, in general, needed:

(1) when failure or deterioration of control could result in a serious health effect;
(2) when measurement is necessary so as to be sure that a MEL or OES, or any self-imposed working standard, is not exceeded; or
(3) as an additional check on the effectiveness of any control, and always in the case of a substance specified in Schedule 4 (vinyl chloride monomer and vapour or spray given off from vessels in which an electrolytic chromium process is carried on, except trivalent chromium).

The HSE provides guidance on suitable sampling techniques and methods of analysis. A further source mentioned in the Code of Practice is occupational health text books.

Records must be kept. Where the record represents the personal exposure of identifiable employees, it should be kept for forty years, and in any other case for at least five years. The records should be kept in such a way that the results can be compared with any health records under Reg. 11, which has already been dealt with in Chapter 4.

Reg. 12 directs an employer who undertakes work which may expose any of his employees to hazardous substances to provide suitable and sufficient information, instruction and training to enable him to know:

(1) the nature and degree of the risks to his health, including any factors which may influence it, e.g. smoking, and
(2) the precautions which should be taken, including the role of health surveillance.

In particular, the employer has a specific statutory obligation to give to his employees and to safety representatives information on the results of any monitoring of exposure at the workplace and of any case where the MEL was exceeded. He must also give information on the collective results (without any individual being identified) of any health surveillance undertaken in accordance with Reg. 11.

Reg. 13 and Schedule 6 require the employer in some circumstances to notify public officials, like a harbour authority or the police, of intended fumigations with hydrogen cyanide, ethylene oxide, phosphine or methyl bromide. Suitable warning notices must also be placed at all points of reasonable access to the premises to be fumigated. The regulation does not apply to fumigations carried out for research.

COSHH was amended in 1991 and 1992. The 1991 Regulations prohibit the importation, supply and use at work of benzene. The 1992 Regulations implement the EC carcinogens directive. They incorporate a new definition of carcinogen. Both sets of regulations make changes in exposure limits. Health records must now be kept for 40 years.

5.12 Noise at Work Regulations 1989

Noise-induced hearing loss has been well known for over a century, but only recently has legislation been enacted. The mushrooming of claims against employers for occupational deafness has made them more conscious of the risks and more inclined to introduce regular audiometric testing, though as yet there is no legal obligation on the employer to provide such tests. The European Commission argues that in failing to provide for compulsory testing the UK has not implemented the EC noise directive and is challenging the UK Government in the European Court.

By the regulations employers are required to make a formal noise assessment where employees are likely to be exposed to 85 decibels (dB) (first action level) or above or to peak action level (200 pascals) or above. Records must be kept. Every employer shall reduce the risk of damage to the hearing of his employees from exposure to noise to the lowest level reasonably practicable (Reg. 6). Where the employee is likely to be exposed to the second action level (90 dB) or above or to the peak action level or above, the employer must reduce exposure so far as is reasonably practicable by methods other than the provision of personal ear protectors. These are appropriate between 85–90 dB and should be provided at the employee's request, but can also be used at higher levels to reduce the exposure to below 90 dB. Ear protection zones should be established with signs indicating that hearing protection should be worn.

5.13   The 1992 regulations

The changes in Europe brought about by the Single European Act gave considerable impetus to health and safety law. A framework directive and six 'daughter' directives were agreed in 1989 and 1990 and these were enacted into UK law as regulations made under the Health and Safety at Work Act. They are the Management of Health and Safety at Work Regulations 1992, the Workplace (Health, Safety and Welfare) Regulations 1992, the Provision and Use of Work Equipment Regulations (PUWER) 1992, the Personal Protective Equipment Regulations 1992, the Manual Handling Operations Regulations 1992 and the Display Screen Equipment Regulations 1992.

All these provisions came into force to some extent on 1 January 1993, but transitional provisions are likely to cause considerable confusion for some years. For example, the new regulations about display screen equipment apply immediately to workstations put into service after 1 January 1993, but not to pre-existing workstations. The latter need only comply with the regulations after 31

December 1996. If new display screen equipment is put into a pre-1993 workstation it becomes a new workstation, but changing a chair or desk does not activate the regulations, except with regard to the new piece of furniture. This is not to say that 'old' workstations are outside the law. The employer will have to comply with the general duty to do that which is reasonably practicable to protect his employees against injury, and he will be bound immediately by the new Management regulations to assess and record risks to the health and safety of employees and others who work with display screen equipment.

For lawyers, these regulations represent a major revision, because they repeal many sections of such legislation as the Factories Act, the Offices, Shops and Railway Premises Act, the Abrasive Wheels Regulations, the Protection of Eyes Regulations and the Woodworking Machines Regulations. Transitional provisions provide that these laws will remain in force side by side with the new regulations for a number of years. Since all except the Management regulations give rise to civil liability, the action for damages for breach of statutory duty has been significantly affected. The increase in the numbers of regulations, Approved Codes of Practice and HSE Guidance will assist employee plaintiffs in their search to make employers liable.

Since the new laws enact EC directives, the courts in interpreting them must strive to implement the intentions of the Council of Ministers. Also, it may be possible to argue that the regulations are out of line with the directive in some particular respect, thus allowing a possible appeal to the European Court.

5.14   Management of Health and Safety at Work Regulations 1992 and ACOP

These came into force on 1 January 1993 and are of immediate application. They lay down broad general principles which apply to virtually all places of work except sea-going ships. No civil action for breach of statutory duty lies in respect of these regulations, but it is likely that they will be regarded by courts as setting new standards of care in common law negligence. The central provision is Reg. 3. Every employer shall make a suitable and sufficient assessment of (a) the risks to the health and safety of his employees to which they are exposed while they are at work; and (b) the risks to the health and safety of persons not in his employment arising out of or in connection with the conduct by him of his undertaking, for the purposes of identifying the measures he needs to take to comply with the requirements and prohibitions imposed upon him by or under the relevant statutory provisions. An employer who employs five or more employees must record (this can be computerised) the significant findings of the assessment and any group of his employees identified by it as being especially at risk.

Risk assessment may already be familiar from compliance with the COSHH Regulations and the Noise at Work Regulations. There are no fixed rules about how it should be undertaken. Paragraph 16 of the Approved Code of Practice sets out the most important factors. A risk assessment should:

(1) ensure that all relevant risks and hazards are addressed (they should not be obscured by an excess of information or by concentrating on trivial risks);
(2) address what actually happens in the workplace or during the work activity (remembering that what happens in practice may differ from the works manual and not forgetting the non-routine operations such as maintenance);
(3) ensure that all groups of employees and others who might be affected are considered (not forgetting cleaners, security staff, visitors, etc.);
(4) identify workers who might be particularly at risk, such as young workers, disabled staff, etc.;
(5) take account of existing protective and precautionary measures.

Every employer shall make and give effect to such arrangements as are appropriate, having regard to the nature of his activities and the size of his undertaking, for the effective planning, organisation, control, monitoring, and review of the preventive and protective measures (Reg. 4). Every employer shall ensure that his employees are provided with such health surveillance as is appropriate having regard to the risks to their health and safety which are identified by the assessment (Reg. 5).

The employer must appoint one or more competent persons to assist him (Reg. 6). This provision puts more pressure than before on employers to appoint safety advisers and occupational health personnel with recognised specialist qualifications (though there is still no statutory requirement). Employers must establish procedures to cope with emergency situations (e.g. fire drills) (Reg. 7). Every employer shall provide his employees with comprehensible and relevant information (Reg. 8). There is now a statutory duty on the employer to tell the employees about any special health risks involved in their work.

Where two or more employers share a workplace each employer has a duty to co-operate with the other and to co-ordinate the measures he takes with those the other employers are taking (Reg. 9). An employer in control of a site must give health and safety information to the employees of contractors (Reg. 10). Where there is no controlling employer, the employers and self-employed persons present should agree such joint arrangements, such as appointing a health and safety co-ordinator, as are needed to comply with the regulations.

Employees must be provided with adequate health and safety training during working hours (Reg. 11). Employees must obey the employer's instructions as to the use of equipment, safety devices, etc. and must inform the employer of any work situation which they would reasonably consider a serious and immediate danger and any matter which they would reasonably consider represented a shortcoming in the employer's protection arrangements (Reg. 12). Reg. 13 deals with temporary workers (defined as those working under a fixed term contract). It obliges the employer to provide information about any special occupational qualifications or skills needed to do the job safely and also any health surveillance required by statute.

5.15 *The Workplace (Health, Safety and Welfare) Regulations 1992 and ACOP*

These came into force on 1 January 1993 as respects new workplaces or new parts of workplaces. They supersede many provisions of the Factories Act and the Offices, Shops and Railway Premises Act, but existing workplaces will only be covered from 1 January 1996. The exception is Reg. 17 which imposes an immediate obligation to provide traffic (including pedestrian) routes in all workplaces which are, so far as is reasonably practicable, suitable for the persons using them, sufficient in number, in suitable positions and of sufficient size.

These regulations affect all places of work except ships, building operations and works of engineering construction and workplaces undertaking exploration or extraction of mineral resources (mines, quarries, oil rigs). Regulations already exist in relation to those places of work. Moving vehicles are covered while stationary inside the workplace (e.g. while being unloaded). They include provisions about ventilation, temperature, lighting, cleanliness, floors, windows, doors, lavatories and washing facilities. Paragraph 239 of the Approved Code of Practice states that rest areas and rest rooms should be arranged to enable employees to use them without experiencing discomfort from tobacco smoke. Methods of achieving this include the provision of separate rooms or areas for smokers and non-smokers, or the prohibition of smoking in rest areas and rest rooms.

5.16 Provision and Use of Work Equipment Regulations (PUWER) 1992

These came into force on 1 January 1993 as respects work equipment first provided for use in the workplace after that date. Existing equipment is not affected until 1 January 1997. It is the date of supply to the employee which is important, not the date of manufacture. An employer who on 2 January 1993 for the first time provides a second-hand van for his employee to use at work must observe the regulations. The same applies to equipment hired or leased after the due date. But an employer who moves existing equipment from one site to another will not have to comply immediately, nor will equipment which was delivered to the place of work before the due date, but only put into use after it, be affected by the new laws.

Equipment means 'any machinery, appliance, apparatus or tool, and any assembly of components which, in order to achieve a common end, are arranged and controlled so that they function as a whole', e.g. combine harvester, automatic car wash, computer, crane, hammer, linear accelerator. It does not include livestock, substances, structural items, or privately owned vehicles. Only workers on sea-going ships are excluded. The Regulations cover not only the situation where the employer provides work equipment for his employee, but also apply to equipment provided by the employee for use at work (other than a private car).

Every employer shall ensure that work equipment is so constructed or adapted as to be suitable for the purpose for which it is used or provided (Reg. 5). 'Suitable' means suitable in any respect which it is reasonably foreseeable will affect the health or safety of any person. Every employer shall ensure that work equipment is maintained in an efficient state, in efficient working order and in good repair (Reg. 6). The HSE Guidance indicates that efficient means in relation to health and safety, not productivity. Every employer shall ensure that all persons who use work equipment have available to them adequate health and safety information and, where appropriate, written instructions pertaining to the use of the work equipment (Reg. 8). Every employer shall ensure that all persons who use work equipment have received adequate training for purposes of health and safety (Reg. 9).

Work equipment will have to comply with EC product standards (Reg. 10) . This part of the legislation applies to brand new equipment provided after 31 December 1992 (second-hand equipment is included only if imported from outside the EC). There is a growing number of EC directives setting out essential safety requirements which manufacturers must meet before putting the goods on to the market. Compliance with the EC standard is demonstrated by affixing a CE mark to the goods. At present not all work equipment is covered by a product directive, but one of the most significant relevant directives is the Machinery Directive, enacted here as the Supply of Machinery (Safety) Regulations 1992. The enforcement of the regulations in respect of machinery used at work is in the hands of the HSE. The manufacturer has the option of complying with the EC standards or the national standards until 1 January 1995, when the EC standards become mandatory.

PUWER repeals s. 14 Factories Act 1961 which made it an offence of strict liability to fail to guard dangerous machinery. The new law is Reg. 11. Every employer shall ensure that measures are taken which are effective to prevent access to any dangerous part of machinery or to any rotating stock bar or to stop the movement of any dangerous part before any part of a person enters the danger zone. This is to be done by the provision of guards where it is practicable (a higher standard than reasonably practicable, but not as high as strict liability) to do so. Further regulations cover high and low temperatures, controls, isolation from sources of energy, lighting, maintenance and warnings.

5.17 *Personal Protective Equipment Regulations 1992*

These regulations, which came into force on 1 January 1993, repeal the Protection of Eyes Regulations 1974. There are no transitional provisions. Where there is already specific legislation, e.g. lead, asbestos, noise, the regulations will not apply. Personal protective equipment (PPE) means all equipment (including clothing affording protection against the weather) which is intended to be worn or held by a person at work and which protects him against one or more risks to his health or safety, and any addition or accessory designed to meet that objective. It does not include uniforms, offensive weapons, signalling devices, seat belts and crash helmets worn on the public highway, sports equipment.

Every employer shall ensure that suitable PPE is provided to his employees who may be exposed to a risk to their health or safety while at work except where and to the extent that such risk has been adequately controlled by other means which are equally or more effective (Reg. 4). The HSE guidance indicates that PPE is always to be regarded as a last resort: engineering controls should be considered first. Since there is a specific requirement to provide PPE for any work where it is necessary, the employer cannot levy any charge on the employee (s. 9 HSWA 1974). The equipment provided must comply with EC standards (Personal Protective Equipment (Safety) Regulations 1992). It must be regularly checked and maintained, and employees must be given instruction and training in its use. Employees must use PPE provided and must report any loss of or obvious defect in the equipment.

5.18   Manual Handling Operations Regulations 1992

These came into force on 1 January 1993. They repeal s. 72 of the Factories Act 1961. Manual handling means any transporting or supporting of a load (including the lifting, putting down, pushing, pulling, carrying or moving thereof) by hand or by bodily force. The core provision is Reg. 4. Each employer shall, so far as is reasonably practicable, avoid the need for his employees to undertake any manual handling operations at work which involve a risk of their being injured. If this is not possible, the employer must make a suitable and sufficient assessment and reduce the risk of injury to the lowest extent which is reasonably practicable. There are four main factors: the load, the working environment, the task and individual capability. More than a quarter of reported industrial accidents are associated with manual handling, most commonly a strain of the back. The HSE Guidance stresses an ergonomic approach – in other words, fitting the job to the worker and not the worker to the job.

The regulations do not specify maximum weights, nor is there any emphasis on the relative strengths of men and women, young and old. Occupational health nurses are specifically mentioned in the Guidance as having the skills both to assess the worker's capabilities and to identify high risk activities. Allowance should be made for pregnancy where the employer can reasonably be expected to be aware of it. Hormonal changes can affect the ligaments and postural problems may increase as the pregnancy progresses. Particular care should also be taken for women lifting loads within three months of giving birth. Medical advice should be sought where an employee's health problem may make him or her particularly vulnerable.

Each employee while at work shall make full and proper use of any system of work provided by his employer in compliance with the regulations (Reg. 5).

5.19 Health and Safety (Display Screen Equipment Regulations) 1992

These regulations came into force on 1 January 1993, but do not apply to existing workstations in service prior to that date until 31 December 1996. This is the first legislation in this growing area of employment. The main hazards so far identified are stress, eye strain and upper limb disorders. There is no evidence to substantiate fears of over-exposure to radiation or higher levels of miscarriage and birth defect. The regulations only cover those who habitually use display screen equipment as a significant part of normal work. This is a question of fact and degree. There are two categories of workers protected: users (employees) and operators (self-employed persons). Employees who work at home are included, whether or not the workstation has been provided by the employer, as are those who work at another employer's workstation.

Every employer shall perform a suitable and sufficient analysis of workstations for the purpose of assessing risks to health and safety (Reg. 2). Workstations must comply with the detailed requirements of the regulations (Reg. 3). Every employer shall so plan the activities of users at work in his undertaking that their daily work on display screen equipment is periodically interrupted by breaks and changes of activity (Reg. 4). The employer must provide training for employees and information for employees and self-employed (Reg. 6 and 7).

The OH professional is likely to become involved with eye and eyesight testing. Only users, i.e. employees, are entitled to tests by competent persons which must be provided free by the employer on request. Both registered ophthalmic opticians and registered medical practitioners with suitable qualifications are competent to undertake these tests. Vision screening tests may also be offered. Tests may be requested at the beginning of employment and thereafter should be repeated as the doctor or optometrist advises. Any evidence of injury or disease should be referred to the general practitioner, with the employee's consent.

If it is found that the user needs special glasses, because normal spectacles or contact lenses will not serve, the employer must provide these free of charge. Less than 10 per cent of the population will need special corrective appliances for display screen work.

5.20 Protection of the environment

The OH professional and the Health and Safety Executive have up to now been seen as the protectors of the workers. In the last decade of the twentieth century there is a growing awareness of the need to protect the environment and the public outside the factory gates. The White Paper: *This Common Inheritance*, published in 1990, was the first comprehensive statement by the British Government of its policy on issues affecting the environment. The Secretary of State for the Environment has general responsibility for co-ordinating the work of the Government on environmental pollution. In addition, local authorities and a wide range of voluntary organisations are involved in environmental protection. In England and Wales Her Majesty's Inspectorate of Pollution (HMIP) has an important role in the control of releases to land, air and water from certain industrial processes through the mechanism of integrated pollution control. HMIP was formed in 1987 by the merger of three existing Inspectorates: the Industrial Air Pollution Inspectorate (formerly part of the Health and Safety Executive), the Hazardous Waste Inspectorate and the Radiochemical Inspectorate, together with a new Water Pollution Inspectorate. Its main function is to implement the Environmental Protection Act 1990 through the authorisation and monitoring of prescribed processes. It has power to prosecute operators of processes which breach release limits set in their authorisations and, by analogy with the Health and Safety at Work Act, to issue enforcement notices (where the operator is in breach of his authorisation) and prohibition notices (where there is an imminent risk of serious pollution of the environment). Environmental health officers of local authorities have similar powers. In order to obtain authorisation operators will have to use best available techniques not entailing excessive cost. If there is one technology which reduces emission of polluting substances by 90 per cent and another which reduces the emissions by 95 per cent, but at four times the cost, it may be a proper judgment to hold that because of the small benefit and the great cost the second technology would entail excessive cost. If, on the other hand, the emissions are particularly dangerous, it may be proper to judge that the additional cost is not excessive.

Enforcing authorities must establish and maintain a public register of applications for authorisation, authorisations, enforcement and prohibition notices and criminal convictions.

The 1990 Act imposes only criminal sanctions. Strict civil liability for environmental damage and pollution is likely to be introduced when the European Civil Liability Waste Directive is implemented.

Chapter 6

The Law of Compensation: Welfare Benefits

6.1 The purposes of a system of compensation

The direct enforcement of safety regulations is in the hands of the various inspectorates who in the last resort can ask for the imposition of criminal sanctions by the magistrates or the Crown Court. Where no-one has suffered injury from a breach of the law, that will be the end of the matter, but for an individual injured by an accident at work or suffering from an industrial disease compensation is probably far more important. Parliament has for nearly a century provided for a system of no-fault compensation for employees injured at work. The introduction of this system in 1897 had a dramatic effect in reducing the number of factory accidents, because employers became aware of the cost of accidents and the fact that often it was cheaper to take safety precautions. In addition, it is arguable that the fear of legal liability to pay damages in the civil courts, especially if numbers of employees are involved, operates as another inducement to the employer to take care. Of course, it is not the employer but the DSS or his private insurance company who actually pays the compensation. The Robens Committee in 1972 examined the interplay between compensation and accident prevention. It was represented to them that insurance is basically inimical to accident prevention because the careful employer is subsidising the negligent organisation. They recommended that the statutory industrial injuries scheme should provide for differential rates of employers' contributions based on claims experience, as in France and Germany, but this has never been implemented.

As for the tort-based action for damages, because this in most cases requires the injured plaintiff to establish negligence, a failure to take reasonable care by the employer, the right to compensation depends partly on luck and having a good lawyer. It is also very expensive to administer, so that each pound in compensation probably requires 85p in legal costs to procure. The fear of legal liability may make it difficult to investigate an incident and take steps to prevent its recurrence, because no-one will talk freely. Employers are obliged by law to insure with private insurance companies against liability for injury to their employees (Employers' Liability (Compulsory Insurance) Act 1969), and although the insurance companies exercise some control, as by inspecting fire precautions and equipment, the careless employer does not suffer the direct cost of compensating his employee. 'High risk' employers have to pay higher premiums than others, but these will be based as much on the general experience of the industry as on the safety record of that company.

If the system of compensation cannot be justified by its preventive effects, the only rationale for its existence is to provide the worker who has been injured in the cause of furthering his employer's business with enough money to make up to him for the less comfortable life he is now living, or to his family for the loss of their breadwinner. Social justice demands that a worker who has undertaken risks for the economic benefit of the community should be cushioned against the consequences of industrial accidents and disease. The fact that many workers pay for their own insurance by reduced wages is an unavoidable consequence of the relative economic strengths of employer and employee.

The Workmen's Compensation Act of 1897 was designed to secure that compensation for industrial accidents should be based on the principle that any employee injured by an 'accident arising out of and in the course of employment' would have an automatic claim against his employer. In 1906, the scheme was extended to disabilities caused by a work-related disease which had been prescribed for that particular employment. The no-fault element is still the major difference between the industrial injuries insurance scheme and the tort-based system of damages. In the former the employee does not have to prove that his injury was caused by anyone's fault: it is enough that he has suffered, whereas in the latter he must establish some fault on the part of a defendant, usually his employer. Until 1946 workmen's compensation was financed in the main by employers paying premiums to private insurance companies. However, the growth from 1911 onwards of a national insurance scheme financed through contributions levied by central government and administered by civil servants meant that those who had not suffered industrial injury but were unable to work because of illness which was not job-related had also become entitled to some benefits.

Beveridge, the architect of the new Welfare State, thought that there were weighty arguments in favour of the merger of the two systems:

> 'If a workman loses his leg in an accident, his needs are the same whether the accident occurred in a factory or the street; if he is killed, the needs of his widow and other dependants are the same, however the death occurred.'

He was persuaded, however that the industrial preference should continue. First, many industries vital to the community were dangerous, yet workers must not be deterred from entering them. Secondly, he felt that a man disabled during the course of his employment had, like a soldier, been disabled while working under orders. Thirdly, he thought that the only justification for limiting the employer's common law liability to damage caused by negligence was that there was a parallel system of no-fault compensation.

A separate and more favourable industrial injuries scheme was retained in the National Insurance (Industrial Injuries) Act 1946, though it was now removed from the private insurance companies and administered by the State. Since then, industrial injuries benefits have gradually been reduced and other social security benefits increased, so that now the industrial preference has been eroded but not abandoned. Consolidation of all the principal social security statutes took place in 1992 in the Social Security Contributions and Benefits Act and the Social Security Administration Act.

6.2   An outline of the scheme

'Employed earners' are covered from their first day at work, because this is not a benefit which depends on contributions by the employee. The statutory definition of an employed earner is 'a person who is gainfully employed in Great Britain either under a contract of service, or in an office with emoluments chargeable to income tax under Schedule E'. A trade union shop steward who sustained an injury at a college where he was attending a training course was an office-holder. Most self-employed workers are outside the industrial injuries scheme, and this can be a serious disadvantage for, for example, construction workers 'on the lump'.

The regulations extend the benefit of industrial injuries insurance to apprentices.

Employees working in the National Health Service are excluded from the general scheme, but protected by the parallel and more generous NHS Injury Benefits Scheme. This provides temporary or permanent benefits for all NHS employees who lose remuneration because of an injury or disease attributable to their NHS employment. The scheme is also available to medical and dental practitioners. It must be established that the injury or disease was acquired during the course of work. An employee who suffers a needle-stick injury is advised to have a serum sample taken at the time of injury and follow-up samples at appropriate intervals in case of infection by, for example, hepatitis B. For those having to give up their employment, the scheme provides a guaranteed income of 85 per cent of pre-injury NHS earnings. There is also a right to a lump sum and dependants' benefits where death occurs. The scheme is administered within the NHS.

Workers on British ships and aircraft and on offshore oil rigs are covered, as during their first year abroad are workers employed in a foreign country by an employer with a place of business here. Reciprocal agreements with foreign countries often qualify those who work abroad for protection under the foreign system.

They can claim if they either

(1) suffer a personal injury caused by accident arising out of or in the course of employment, or

(2) suffer from a 'prescribed disease', that is one designated by the Secretary of State, usually after consultation with the Industrial Injuries Advisory Council (IIAC), as a special risk for particular occupations, e.g. viral hepatitis for those working in contact with human blood or human blood products, or tenosynovitis for manual labourers or those in jobs involving frequent or repeated movements of the hand or wrist.

It is presumed that the disease was in fact caused by the performance of the prescribed occupation, unless the adjudication officer proves to the contrary. So far, the Government has not accepted the proposal of the IIAC that claims should be allowed where the claimant is able to show that his disease was caused by his work even though not yet prescribed for it by the Secretary of State (often called a case of 'individual proof'). However, in 1991 occupational asthma was prescribed not only for jobs involving contact with such well-established sensitisers as isocyanates and laboratory animals but also for work with 'any other sensitising agent'. Claimants must name the suspected substance and must be examined by a Medical Board. In the first year of operation 129 claims were successful under this head. Substances found guilty of causing occupational asthma included resins, chrome, nickel and icing sugar!

The Industrial Injuries Advisory Council includes representatives of employers and unions and doctors, epidemiologists, toxicologists and lawyers. The suspect disease must be proved with reasonable certainty to be attributable to work and not a risk common to all persons. Epidemiological evidence is therefore of great importance. There are four categories of occupational disease:

(1) Physical (e.g. cramp of the hand or forearm due to repetitive movements).
(2) Biological (e.g. viral hepatitis, Q fever).
(3) Chemical (e.g. lead poisoning, certain cancers caused by exposure to chemicals).
(4) Miscellaneous conditions (e.g. pneumoconiosis, mesothelioma, non-infective dermatitis).

In recent years an 'intermediate' form of prescription has been adopted, whereby a disease is listed subject to certain conditions. For example, compensation for occupational deafness will only be paid if the claimant can show hearing loss of at least 50 dB in each ear, being the average of hearing losses at 1, 2 and 3 kHz frequencies, and being due in the case of at least one ear to occupational noise. The claimant must have been employed for a minimum aggregate of 10 years in one or more of the qualifying occupations, the claim must be made within five years of the claimant last working in one of the qualifying occupations and the claimant must show that the hearing loss in at least one ear is due to noise at work. 'Noisy hobbies such as musketry and listening to amplified pop music should not be overlooked': (*DSS Industrial Injuries Handbook for Adjudicating Medical Authorities.*)

An amendment in 1989 removed the requirement that hearing loss should be measured by pure tone audiometry as opposed to evoked response audiometry or any other test.

Vibration white finger (Raynaud's phenomenon) was not prescribed until 1985. Three or more fingers must be affected throughout the year for a claim to be made. In 1993 carpal tunnel syndrome became a prescribed disease for work using hand-held vibrating tools. Chronic bronchitis and emphysema were in 1993 prescribed for miners after many years of debate.

6.3 The benefits available

When a worker is first struck down by accident or disease, he needs a guaranteed income to pay his household expenses, mortgage and so on. Here there is now no industrial preference. All employees are entitled to statutory sick pay from their employer, whatever the cause of their absence from work. The majority will also be entitled to additional contractual benefits depending on the terms of the contract of employment, which may be more generous to those absent through industrial disease or injury.

The Social Security Act 1986 abolished the death benefit whereby dependants of a worker killed by a work-related disease or injury could claim benefit. What is left is a scheme to provide for disablement. It is argued that disabled people should receive similar benefits whatever the cause of their disability, but the IIAC's opinion is that 'the only acceptable reason for, and method of, abolishing the industrial preference would be by extending industrial injuries-type benefits to all disabled people'.

This is also the view of the TUC and, currently, of the Government which in the late 1980s was faced with the alternative either of levelling down industrial injuries benefits or extending them to all disabled people at a cost of over £3,000 million. The abolition of the reduced earnings allowance for those disabled after 1990 was a further step in the reduction of industrial benefits. The Social Security Contributions and Benefits Act 1992 contains separate regimes for the disabled and the industrially disabled. Appeals by the non-industrially disabled are referred to Disability Appeal Tribunals.

Another important change brought about by the 1986 Act was the abolition of lump sum payments for small handicaps. The philosophy behind the scheme is now to maintain the living standards of those injured at work rather than to give an award of 'damages', as in the days of workmen's compensation. If the employee wishes to obtain a lump sum he will have to pursue a civil action against the person allegedly responsible for his injury.

There is now only one benefit: a disablement pension. The additional reduced earnings allowance (known as special hardship allowance up to 1986) has been abolished in respect of losses of faculty resulting from an accident or the onset of a prescribed disease on or after 1 October 1990. Reduced earnings allowance is still being paid to those disabled before October 1990 up to the date they reach pensionable age. All payments under the scheme are tax-free.

The disablement pension compensates for injury to the employee's physical or mental health caused by work-related accident or disease. It is a regular payment which continues as long as the disablement lasts, in many cases up to the claimant's death. The pension is payable whether or not the claimant is employed and whether or not his earnings are reduced. To receive a pension, the claimant must be assessed by a medical panel as at least 14 per cent disabled, except in the case of those suffering from pneumoconiosis, byssinosis or diffuse mesothelioma. The amount of the pension depends on the degree of disablement. Total loss of hearing or amputation of both hands can amount to 100 per cent disablement: it is not necessary for the worker to be completely bedridden. The assessment of the degree of disability is made by doctors, but there are guidelines laid down in regulations: e.g. loss of whole index finger – 14 per cent; loss of thumb – 30 per cent; loss of both hands – 100 per cent.

Supplements are payable to those who are 100 per cent disabled and need constant care. These are known as the constant attendance allowance and exceptionally severe disablement allowance.

6.4 Making a claim

Application for industrial injuries benefit must in the first place be made to the local adjudication officer (AO) at the DSS offices. There is no entitlement for the first 15 weeks following the accident or onset of the prescribed disease. Disablement must be assessed as at least 14 per cent for the disablement pension (except for sufferers from pneumoconiosis, byssinosis or diffuse mesothelioma where disablement must be assessed as at least 1 per cent. The AO has power to decide all relevant issues except whether the claimant is an employed earner (to be decided by the Secretary of State) and disablement questions (to be determined by medical authorities).

If benefit is refused, an appeal can be made to a Social Security Appeal Tribunal (SSAT). The tribunal consists of a legally qualified chairman and two 'wing members', lay persons representing the community. If the SSAT turns down the claim, a further appeal can be taken to a Social Security Commissioner who is a barrister or solicitor, but only on a point of law. A trade union can appeal on behalf of a claimant who is a member. The appeal may be decided on the papers, or an oral hearing may be held, in London, Edinburgh or Belfast. Legal aid is not available to pay for legal representation, but expenses are payable. Further appeals on law can be made to the Court of Appeal, but only with the leave of the Commissioner or the Court of Appeal, and ultimately to the House of Lords.

In deciding on a claim for industrial injuries benefit, medical questions will arise, first as to the correct diagnosis, and second as to the degree of disablement. These are almost invariably referred to an adjudicating medical authority (AMA). An AMA is either a single doctor or two or more doctors drawn from a panel appointed by the Secretary of State, usually composed of general practitioners. Specially qualified medical boards are appointed to deal with pneumoconiosis and ten other diseases (byssinosis, diffuse meso-thelioma, asthma, lung cancer, pleural thickening, poisoning by nitrous oxide, beryllium, or cadmium, nickel cancer and alveolitis). The AMA takes a full history from the claimant who is usually allowed to be accompanied by a representative (unless the doctor refuses). The doctor then physically examines the claimant, usually in private. If the case has been referred to two doctors and they disagree it will have to be submitted to another board with three members. The regulations provide that everyone who claims industrial injury benefit must, if required to do so, submit to a medical examination and to such medical treatment as is rec-ommended for his industrial injury, on pain of disqualification from benefit.

If the claimant (or the Secretary of State) wishes to challenge the medical assessment he may do so in writing to a Medical Appeal Tribunal (MAT). This consists of a legally qualified chairman and two doctors, usually consultants, not previously involved with the claimant. The MAT examines the claimant and holds an oral hearing at which the claimant can put his case: expenses are paid but legal aid is not available. The MATs hear five times as many industrial injuries claims as SSATs and something like 45 per cent of appeals by claimants are successful. If the claimant wishes his hospital or other medical records to be put in evidence he must ask the tribunal to obtain them. The physician giving records to the tribunal should first obtain written consent from the claimant. The tribunal must give the claimant a précis of the medical reports if relied upon, unless knowledge of them would harm the claimant's physical or mental health.

Appeal from an MAT lies on a point of law only to a Social Security Commissioner. Difficulty has arisen in the past from conflicting decisions of an SSAT (known as an adjudicating authority) and an MAT (known as a medical authority). It is now possible for an AO to decide that personal injury has been caused by accident, for example a myocardial infarction caused by lifting, and for the doctors then to overrule him by resolving that the claimant is not entitled to a pension because his condition entirely predates the alleged accident.. Questions about the degree of disablement, however, are left to the medical authorities alone.

6.5 Review of decisions

Sometimes, the claimant wants his case to be looked at again rather than appealed to a higher level. He may argue that the authorities have made a mistake of fact or that fresh evidence is now available. The new evidence must be such that it could not be obtained with reasonable diligence for use at the trial. The claimant may ask for a review of his assessment on the ground that sequelae have now become apparent, under the heading of unforeseen aggravation, but it is not possible to reopen a final (as opposed to a provisional) assessment because of an unforeseen improvement in the claimant's condition, for otherwise claimants would have no incentive to achieve rehabilitation. The Social Security Act 1986 has introduced a new provision allowing a review at any time of a decision of an AMA by that practitioner if he is satisfied that he has made a mistake of law.

6.6 Compensation for accidents

The claimant must show

(1) personal injury;
(2) which has been caused by accident;
(3) which has arisen out of and in the course of employment.

An accident, a word notoriously difficult to define, is an unexpected event, and in industrial injury cases may even include the murder of a schoolteacher by his pupils in the playground (*Trim School* v. *Kelly* (1914)). Accident includes a slipped disc caused by heavy lifting, even if the claimant realised at the time that his back might suffer. It differs from a disease in that it occurs over a short period and not by a continuous process, but this may be a fine distinction in practice. Silicosis caused by exposure to dust over many years is not an accidental injury, nor is deafness gradually induced by the noise of a mechanical saw used for cutting up meat, whereas a psychoneurotic illness caused by irregular explosive reports from a machine has been held to be accidental on the argument that each separate explosion was an accident. In 1986 a Scottish civil servant who had been off work for 19 months claimed industrial injury benefit on the basis that he had developed acute anxiety tension due to various strains at work, including diminishing prospects of promotion and transfers to different departments. It was held by the Social Security Commissioner that strain arising from uncongenial working conditions could not be described as caused by an accident (*Fraser* v. *Secretary of State for Social Services*). However, it was subsequently held by the English courts that a civil servant who had suffered a mental breakdown through overwork was entitled to benefits under the special and more generous terms of the civil service pension scheme which includes 'injury or disease attributable to or caused by' his work, but only if he could establish by medical evidence that the illness was caused by the pressure of work combined with his personality (*R* v. *Minister for the Civil Service, ex parte Petch* (1987)).

A health professional who contracts the HIV virus through a sudden event, for example a needle-stick injury or a violent attack from a patient, would be able to claim compensation for accidental injury, despite the absence of AIDS from the list of prescribed diseases. In *Clay* v. *Social Security Commissioner* (1990) the claimant, an asthmatic, from time to time suffered painful attacks of nausea, breathlessness and headaches through passive exposure to cigarette smoke at work. It was held that she was entitled to benefit for personal injury caused by accident.

The accident must cause personal injury to either physical or mental health. Damage to property is excluded (even to false teeth or a wooden leg). It is very important that a potential claimant should be certain that a record is kept of industrial accidents, as proof of what occurred. One of the best ways of giving notice is to enter the details in the accident book. In the case of accidents or diseases within the RIDDOR Regulations, notice must be given by the employer to the Health and Safety Executive (HSE) (see Chapter 3).

If the cause of the claimant's injury is outside his employment he will not be entitled to industrial injury benefit. Thus, an employee who suffers from a pre-existing heart condition cannot claim because his heart attack occurs by chance when he is at work. But if the work causes or contributes to the injury in any material degree a claim can be made, so that if there is evidence that the attack was brought on by the employee having to lift a heavy cupboard at work he will be covered. It is irrelevant that an average employee would have had no problems with the cupboard. In addition, if the employee collapses at work and injures himself on his employer's building or machinery, that injury will be deemed an industrial accident.

Difficulties arise over the interpretation of the words 'arising out of and in the course of employment'. They are not confined to the employer's premises during strict working hours, but have to be given a broad common sense interpretation. An accident which occurs in the course of employment is presumed to arise out of it. The Social Security Act specifically provides that accidents which occur in the course of employment and are caused by another person's misconduct, skylarking or negligence, or the behaviour or presence of an animal (including a bird, fish or insect), or are caused by or consist in the employee being struck by any object or by lightning, are to be treated as arising out of employment.

6.7 Accidents on the way to work

'Commuter' accidents are not included in the scheme, unlike in France, Australia, Germany and Sweden, and contrary to the recommendation of the Pearson Commission. Normally a person's employment begins when he arrives at his place of work and ends when the person leaves it. If, however, the travel from one place to another is part of the job, as with a lorry driver or commercial traveller, benefit will be payable. What about the journey from home to the first call of the day and the return home from the last job? If the 'travelling' employee's home is his base from which he works these will be covered. It has been held that a local authority home help injured on her way to her first house of the day was not entitled to claim because the nature of the job is not to travel. The home help travels to duty, not on duty. Contrast with this *Nancollas* v. *Insurance Officer* (1985) in which a disablement resettlement officer employed by the Department of Employment was involved in a road accident on the way from his home to an employment office at which he had arranged to interview a disabled person. He was based at an office in Worthing and made many visits to job centres and private houses in the course of his work. Many of his journeys commenced and terminated at the Worthing office, but in this case, because the appointment was early in the morning, he started from home. The Court of Appeal decided that Nancollas could claim industrial injury benefit because on a common sense reading of the situation he was acting in the course of his employment as he drove to his appointment. This decision shows the unpredictability of the law in this area. The claimant was not based at home, but he was covered on the way to his first appointment because he was regarded as an employee with no precisely definable place of work. Unlike the home help, he travelled to many different destinations.

The law provides that an employee injured in an accident on the way to or from his place of work while a passenger in a vehicle operated by or on behalf of his employer or by arrangement with his employer, not being a public transport service, shall be entitled to industrial injuries benefit. Such travel must be with the express or implied permission of the employer.

6.8 Accidents during breaks from work

Normal and reasonable breaks in work are included in the course of employment as long as the employee remains on his employer's premises. Accidents in the canteen or the toilets are covered as incidental to the employment. But if the employee leaves his place of work, he will not usually be protected, unless he is employed to travel, like a lorry driver stopping at a motorway service station. An employee who disobeys instructions may exclude himself from protection. A dock labourer who used an unattended forklift truck to move an obstacle was denied benefit (*R* v. *Alburqerue ex parte Bresnahan* (1966)). In *R* v. *Industrial Injuries Commissioner ex parte AEU* (1966) an employee waited outside a booth provided by the employer in the factory so that workers could safely smoke during breaks. Because the booth was full, he remained there for five minutes after the tea-break had ended and was hit by a fork-lift truck. The court, somewhat harshly, denied him benefit because he had taken himself outside his employment by overstaying his break. Workers who were injured using occupational health facilities provided by the employer have made successful claims, as did an employee hurt on the way to a mobile X-ray unit parked on his employer's premises.

If the employee arrives early or stays late he will still be covered as long as his presence is reasonably incidental to his employment, but not if it is simply to suit his own convenience.

Employees who are injured away from their employer's premises may claim if acting in the course of employment. Apprentices on day release at a college of education will be covered, and a school teacher injured in Switzerland while accompanying a school party was held entitled to claim. Spare time activities are not included: a policeman injured in a rugby game having volunteered to represent his county force lost his appeal *R* v. *National Insurance Commissioner, ex parte Michael* (1977)).

6.9 Emergencies

If the employee is injured while trying to save people or property in an emergency connected with his job he will usually be covered. A driver who deviated from his journey to obtain medical assistance for a fellow employee made a successful claim. When a dockyard policeman was injured outside the docks in rescuing a child in a runaway pram he received benefit, but an ordinary member of the public acting in the same way would not have been protected, because it would not have been incidental to his employment.

6.10   Compensation for prescribed diseases

An injury caused by a gradual process does not fall within the definition of an accident. The legislation provides that those who gradually over a period contract one of a number of work-related diseases specifically prescribed for particular occupations by the Secretary of State in regulations can also claim benefit. A claimant who is able to prove that his injury has been caused by his job, but cannot bring himself within either the accident or the prescribed disease provisions, will be denied industrial injury benefit. Sometimes claimants try to establish that a non-prescribed disease has been caused by an accident in order to bring it within the 'accident' rules. Infective illnesses may occasionally be included within the category of accident in that there must have been a moment of time in which, in non-medical language, the infection entered the body. Prescribed diseases account for about a quarter of disablement pensions, and pneumoconiosis is by far the most common disease among disablement pensioners.

The claimant must prove:

(1)   employment in an occupation listed in relation to the disease;
(2)   that the disease is prescribed;
(3)   that the disease was caused by the job;
(4)   that he has suffered loss of physical or mental faculty.

There is a general presumption that a prescribed disease is caused by the relevant occupation if the employee worked in it at any time within one month preceding the date of development of the disease. This is varied in the following cases:

(1)   *Occupational deafness.* The claimant must have been employed for an aggregate of at least 10 years in a relevant occupation and have worked in such an occupation within the five years preceding the claim.
(2)   *Tuberculosis.* The claimant must have worked for at least six

weeks in the relevant occupation and have done so within two years before the development of the disease.

(3) *Pneumoconiosis.* The claimant must have worked for at least two years in aggregate in a relevant occupation.

(4) *Byssinosis.* The presumption applies to any employee who has worked in a relevant occupation for any period.

(5) *Inflammation of the nose, throat or mouth, and non-infective dermatitis.* There is no presumption. The employee must prove that his injury is job-related in every case.

(6) *Occupational asthma caused by 'any other sensitising agent'.* The claimant will have to identify the agent and prove that the asthma is job-related.

To rebut these presumptions, the AO will have to prove that the claimant's illness has some other cause, like his living conditions or personal habits. The job does not have to be proved to be the only cause of the disease as long as it is a substantial cause.

6.11　Sequelae

A claimant who contracts a prescribed disease may develop another non-listed condition as a result. This sequela will be treated as an industrial injury even though not prescribed. For example, tenosynovitis may lead to carpal tunnel syndrome and insanity may be a sequela of poisoning by mercury or methyl bromide. Obviously, medical opinion is very important in these cases and the DSS publishes a booklet: *Notes on the Diagnosis of Occupational Diseases.*

6.12　Recrudescence

Where a claimant has suffered from more than one attack of the same illness and that illness is not a chronic one like byssinosis or deafness from which no-one can be expected to recover completely, it will be necessary to decide whether this is the old disease flaring up (recrudescence) or a fresh attack. Dermatitis causes much difficulty. A new attack must give rise to a new claim; recrudescence will be taken into account in reviewing any previous assessment and will also 'back-date' the onset of the disease. There is a rebuttable presumption that an attack occurring during the period of an existing assessment of disablement is a recrudescence of the old disease.

6.13   Pneumoconiosis

There are special rules applying to this disease, involving permanent damage to the lung due to inhalation of mineral dust, and including silicosis and asbestosis. There is a presumption that anyone showing typical symptoms who worked in one of the scheduled occupations for at least two years is entitled to benefit. Most claimants have worked in coal mines or with asbestos. Diagnosis and disablement decisions are made by special medical boards composed of specially qualified doctors. A pension is payable as long as the disability is assessed as at least one per cent, and since the Social Security Act provides that a person found to be suffering from pneumoconiosis shall be deemed to be suffering from loss of faculty of at least one per cent, every case diagnosed is entitled to a payment which in administrative practice is assessed as 10 per cent because of the difficulties of calculation. Appeal lies to a MAT.

When a special board makes a diagnosis of pneumoconiosis it notifies the claimant of whether it is safe to continue at work and, if so, what kind of work. If the claimant gives his consent, the board writes to the employer and asks him to provide the claimant with suitable working conditions. The board also notifies the claimant's GP.

In addition to the statutory scheme, British Coal has a private no-fault compensation agreement with the mining unions, introduced in 1974. This makes lump sum payments on top of industrial injury benefit to coal miners who are in receipt of a DSS disablement pension for pneumoconiosis, or to dependants where the miner has died. Anyone who receives a payment under the scheme must agree to relinquish any right to sue for civil damages at common law. British Nuclear Fuels operate a similar scheme for radiation-induced cancer. The Pneumoconiosis etc. (Workers Compensation) Act 1979 gives workers with specified lung diseases a right to a lump sum from the State, where they cannot recover damages from the employer because he is no longer in business.

6.14   Assessment of disablement

The determination of this question decides whether the claimant is entitled to benefit and, if so, how much. All disablement questions are reserved for the medical authorities and are defined as:

(1)   whether the accident or disease has resulted in a loss of faculty;

(2)   at what percentage the extent of disablement resulting from the loss of faculty is to be assessed and for what period.

6.15   Loss of faculty

The basic pension is to compensate the claimant for the fact that his bodily or mental condition has deteriorated as a result of his industrial injury; he is not the man he once was. It is irrelevant that he has not lost earning capacity. The assessment is to be made, unlike common law damages, without reference to the claimant's personal circumstances, other than age, sex and physical and mental condition. A rich, sedentary restaurant manager receives the same disablement pension for the loss of his hand as a poor waiter who was a keen amateur footballer. However, age and sex are relevant. A physical disfigurement may be assessed more highly in the case of a woman claimant. Stiffness in the joints may not be loss of faculty in a worker of 60, where it would be in one of 18. The claimant must establish on a balance of reasonable probabilities that the loss of faculty resulted from the industrial accident or prescribed disease.

6.16   Disablement

This may or may not flow from a loss of faculty. A small scar in an inconspicuous place is not disabling; an ugly disfigurement of the face causing psychological problems can be a serious disability. In any event, since the abolition of a right to a pension for most disabilities of less than 14 per cent, many minor disabilities fall outside the scheme. A disability has been defined as an 'inability to do something which persons of the same age and sex and normal physical and mental powers can do'.

The medical authorities are given considerable flexibility, but for some injuries they are guided by the statutory tariff. All assessments must be expressed in percentage terms. The assessment is made for the period during which the claimant has suffered and may be expected to suffer from the loss of faculty. If the claimant's condition is such as not to allow of a final assessment, because of possible changes, a provisional assessment is made which can be later reviewed.

6.17 Multiple disabilities

If a man already blind in one eye suffers an injury to his other eye at work, the injury is more disabling than for a fully sighted worker. The two eye injuries are connected conditions because the combination of the two is more disabling than if they were treated in isolation and the percentage disablement will be increased. The regulations provide that the doctors should first assess the combined disability (blindness in one eye plus injury to the other) and then deduct from that figure a percentage to cover the degree of disablement had the industrial accident not occurred (blindness in one eye only). The AMA has a discretion not to reduce an assessment if satisfied that in the circumstances 100 per cent is a reasonable assessment of the extent of disablement from the relevant loss of faculty.

Where the claimant suffers a non-industrial accident after the industrial injury, as for example where a man blinded in one eye at work then suffers an injury to the other eye in a domestic accident, regulations of such complexity that they have been the subject of judicial criticism provide in effect that the combined figure of the two disabilities should be taken together minus the percentage for the non-industrial injury, but only if the industrial injury on its own is assessed at more than 10 per cent disablement. Thus, if the industrial injury counts as 60 per cent, the domestic injury alone as 30 per cent and the connected industrial and non-industrial injuries together as 100 per cent, the claimant's disablement is assessed as 100 minus 30, that is 70 per cent.

Pre-existing conditions which might never have become apparent had the industrial injury not occurred are not to be treated as existing disabilities and are therefore not deductible. The industrial injuries scheme adopts the common law principle that you 'take your victim as you find him'. Thus, in a case in 1981 where a MAT reduced benefit because they found that at the time of the industrial injury the claimant was in any event liable to develop multiple sclerosis, the Commissioner held that no deduction should have been made. In addition, since the claimant is to be compared with an average healthy person of the same age, the percentage disablement should not be reduced because the claimant is already showing signs of normal ageing. A middle-aged worker claiming for a serious back injury should not have his pension reduced because 'at your age you must expect problems with your back'.

6.18 Reduced earnings allowance

This was formerly known as special hardship allowance. It may be additional to the disablement pension. It is payable only if the industrial injury has caused the claimant to lose wages. It has been abolished in respect of injury caused by accidents occurring or the onset of prescribed disease on or after 1 October 1990. The abolition of the allowance was expected to reduce expenditure annually by £130 million by 2001–2.

The claimant must prove:

(1) that he is entitled to a disablement pension, or has been assessed as not less than 1 per cent disabled;
(2) that as a result of the relevant loss of faculty, he is either permanently or temporarily incapable of following his regular occupation and also incapable of following employment of an equivalent standard which is suitable in his case.

6.19 Incapacity for work

Incapacity means that the claimant must show (in cases of pneumo-coniosis the burden is on the insurance officer to show that the claimant is capable) that there is no work which he can reasonably be expected to do. Registration as a disabled person is in itself insufficient evidence of this. Medical evidence is obviously of prime importance. The AO is likely to ask the medical authorities for their opinion, as well as the claimant's own doctor. He may also ask the claimant's employer for a report on the nature of the claimant's normal work and the disablement resettlement officer for an assessment of the claimant's capabilities. If the claimant has returned to work this is good evidence that he is capable, but it is not conclusive. The claimant who finds that he cannot work for as many hours as his contract requires (disregarding voluntary overtime), or perform a task essential to his job, will be incapable.

Where the claimant is found to be only intermittently incapable, he will not be entitled to reduced earnings allowance. The doctors will be asked to decide whether the claimant is likely to remain permanently incapable of following his normal job: if so, he will receive the allowance for as long as he is also incapable of suitable alternative employment. If not (and doctors are reluctant to certify permanent incapacity) the claimant must show that he has been *continuously* incapable of doing his job or a suitable alternative since the end of the 15-week period after the accident or onset of disease. Thus, an employee who returns to work and then finds it is too much for him may have put himself at a disadvantage if he cannot show that he is permanently incapable. To alleviate this potential hardship, the regulations provide that the claimant can return to work 'for the purpose of rehabilitation or training or of ascertaining whether he had recovered from the effects of the relevant injury' and not lose benefit, but only in restricted circumstances. These are that the work is with the approval of the Secretary of State or on the advice of a medical practitioner and that it does not amount in aggregate to more than six months. The claimant's doctor should have sanctioned the return to work; the fact that he did not advise against return may be insufficient.

If the claimant returns to work pending surgical treatment for the effects of his industrial injury, this period may also be discounted under the regulations, as long as a medical practitioner has previously advised surgery, or the claimant while at work was in the process of obtaining such advice (as where he was attending hospital as an outpatient). It is necessary for the claimant to prove that he used reasonable zeal and expedition in trying to secure surgical treatment.

Sometimes the claimant is incapable of work because of some psychological illness. If this has been caused by the industrial injury, as where the employee suffers from claustrophobia as a result of being trapped at work, the phobia is in itself a loss of faculty and should be so diagnosed by the medical authorities. It is irrelevant that the fear is unreasonable. Where the claimant is not suffering from any psychiatric condition, but refuses to return to work or undergo medical treatment which will assist him to do so because he is afraid, he will be denied the allowance unless his fear is found to be reasonable, for example because the suggested treatment carries significant risk of failure.

It is presumed that a claimant who is receiving a disablement pension in respect of pneumoconiosis and who changes his job after having been advised by a special medical board that he should not follow his regular occupation unless he complies with special restrictions relating to the conditions in which he works, will be entitled to REA if he suffers a drop in wages. The presumption may be rebutted if, for example, the claimant was nearing retirement or likely to be made redundant at the time he made the change.

6.20 The regular occupation

What is the claimant's regular occupation? Suppose that the claimant changed his job before the consequences of the industrial disease manifested themselves? Since the REA depends on a comparison between what the claimant would have earned in his regular employment and what he now earns, how can the system cope with a job which has disappeared? A flexible approach has been adopted. As the Commissioner said in a case in 1975:

'[A] fair and realistic view must be taken in the light of the work which the claimant was doing and was required to do and all other circumstances including the claimant's reasons for doing the work he was doing and his intentions for the future.'

In one case, a bus driver agreed to take a job as a storekeeper to see whether he liked it; he then had an industrial accident. It was held that his regular occupation was to drive a bus. But if he had made a definite change in his job, his regular occupation would have been that of storekeeper however short his period of employment in that position. Regulations provide for an exception if the employee has been forced to change his job by the symptoms of industrial disease, though before the official date of onset; here, the regular occupation is that done at the time the symptoms necessitated a change of job.

Sometimes, an employee will be injured at a very young age and will claim benefit for many years. It will be difficult to continue to assess how much the claimant would have been earning if he had not suffered the loss of faculty; if the job has completely disappeared it may be virtually impossible. The Social Security Act 1986 now assists the DSS by providing that after the first assessment of loss of remuneration, all subsequent assessments may be determined by reference to scales of earnings in the relevant industry – that is, they will be index-linked. The regulations also allow for the possibility that the claimant would have gained promotion to a higher grade, as long as he can demonstrate that in general persons in his position are normally promoted. Thus, an injured apprentice can claim on the basis that he would have become a skilled worker.

6.21 Alternative employment

If the claimant cannot do his normal job but is able to undertake another recognised occupation which is of an equivalent standard of remuneration, he will not receive REA, because this benefit is for loss of earning capacity, not loss of a particular job. The alternative employment must be suitable for the claimant in question, taking into account his age, capabilities, education and so on. A job in a different part of the country which would require the claimant to move house is not usually treated as suitable (*R.* v. *Deputy Industrial Injuries Commissioner ex parte Humphreys* (1966)).

6.22   Subsequent ill-health

Lump sum awards of damages at common law allow for the possibility that the plaintiff might not have lived his normal span untouched by accident or disease. REA is regularly reviewed because it is granted only for a period fixed by the authorities. If it is shown that the claimant has become incapable because of an illness unconnected with work, REA will not be renewed. In *R* v. *National Insurance Commissioner ex parte Steel* (1978), Henry Steel was a miner who was diagnosed as suffering from pneumoconiosis in 1959. The medical board told him that he could continue to work in mining as long as he confined his work to approved dust conditions and had regular checks. They continued to give him this advice until 1975 when he gave up work because of breathlessness. This was subsequently diagnosed as caused by hypertension. He was incapable of following his regular occupation, but capable, apart from the hypertension, of following employment of an equivalent standard under approved dust conditions. He was denied special hardship allowance.

Chapter 7

The Law of Compensation: Civil Liability

7.1 The tort action

In 1987, Graham Cook, a man in his thirties, was awarded damages of £850,000 for catastrophic injury suffered when in the course of his employment he placed a batch of scrap batteries in a bath of water and then added nitric acid to extract silver. The batteries had been contaminated with iron pyrites, so that they gave off hydrogen sulphide, a highly toxic gas, when immersed. Mr Cook was totally paralysed, save for some slight movement in his head. He lost the power of speech, but not his intellect, so that he was a conscious prisoner in a useless body, only able to communicate by means of a computer. To obtain his damages, he had pursued a civil action in tort in the High Court in the course of which he had been required to prove that someone had caused his disablement by negligence, a failure to take reasonable care: it was eventually held that his employer was not negligent but that the company which had sent the scrap to be treated was solely responsible, because they should have warned Cook's employers of its contamination (*Cook* v. *Engelhard Industries*). The accident happened in 1982, but judgment was only given in 1987. Even then, it would have been possible for the defendant's insurance company to have taken the case on appeal to the higher courts which would have added to the delay and to the legal costs, and might have meant the reversal of the High Court's decision.

From the standpoint of the injured employee, a more unsatisfactory system of obtaining compensation cannot be imagined. However, Mr Cook was in one sense exceptionally fortunate, in that he was one of the approximately 10 per cent of those injured at work who are successful in a claim for damages for common law negligence. (This was the estimate of the Pearson Commission in 1977.) He would already have been in receipt of social security disablement and reduced earnings benefits. At the time, only half of these payments over five years would have been deductible but

196

now, by the Social Security Administration Act 1992, anyone paying a compensation payment of more than £2,500 must deduct from the damages and pay to the Department of Social Security all the social security payments made to the plaintiff over five years from the accident or onset of the disease.

How does the tort system differ from the industrial injuries scheme? One very significant difference is that damages in tort are paid as a lump sum, rather than as a regular payment. High Court judges have to guess what the future holds, whereas social security tribunals can make provisional assessments and make subsequent variations if circumstances change.

The inadequacy of the rough and ready method of the tort system has now been accepted by the judges in their increasing willingness to approve structured settlements in tort actions. The first such case was *Kelly* v. *Dawes* (1990). The structured settlement must be agreed by both parties: it cannot be imposed by the court. It permits the defendant to pay the bulk of the damages by purchasing an annuity. The Inland Revenue has agreed to treat the annuity payments to the plaintiff as tax-exempt payments of capital. Thus, the amount the defendant has to pay will be reduced, and the plaintiff is assured of a reasonable tax-free income for as long as he or she may live.

Another important point is that the family of a decreased worker can no longer claim for his death under the social security scheme, while those relatives who were financially dependent on him can claim for loss of their breadwinner under the tort system. Tort in most cases requires proof of negligence: this is unnecessary in the no-fault social security system. Claims in tort are referred to the slow, expensive and highly technical County Courts and High Court (though legal aid is available to the less well-off); social security tribunals are quicker, more informal and cheaper (but not covered by legal aid and it cannot be said that the law they administer is free of technicality). Industrial injury compensation is paid from State funds; workers who seek tort damages must establish a case against a defendant who is either wealthy enough to pay or is sufficiently insured against liability. In the end both damages and social security payments are paid from money contributed by the public at large, because all employers are required by law (*Employers' Liability (Compulsory Insurance) Act 1969*) to insure against actions by their employees for work-related injury: they recoup the cost of this insurance from the prices charged for goods and services.

In 1978 the report of the Royal Commission on Civil Liability and

Compensation for Personal Injury (Pearson Commission) examined the effectiveness of the system of compensation for work-
related injury. The members of the Commission were impressed by
the range and levels of benefits available under the social security
system. They thought that the existence of the no-fault scheme
prompted serious consideration of the abolition of the tort action.
The potential risk of civil liability often made it difficult to investigate the causes of an accident, and potential defendants feared
that the introduction of new safety measures might be interpreted
as an admission that previous practices were defective. On the
other hand, the fear of being sued was an incentive to maintain
standards of accident prevention. Only Professor Schilling, an
eminent professor of occupational health, thought that the tort
action should be abolished. He argued that accident prevention is
best secured by strong and effective criminal laws and that compensation should be left to the social security system.

The relative value to the injured worker of the two types of
compensation has been assessed by Richard Lewis (*Compensation
for Industrial Injury*). He writes that the industrial scheme annually
compensates three times as many people as the common law, if
those who are regularly in receipt of disablement benefit are
included. He also points out that the value to an individual of
industrial injury benefits is greater than many suppose, because the
benefits are tax-free income and are regularly updated in line with
inflation. Damages are awarded as a lump sum tax-free payment of
capital, but income tax must be paid on the interest. In addition,
judges do not take future inflation into account in assessing the
amount of the award.

7.2   The employer's duty of care

In cases of industrial injury the plaintiff usually attempts to prove
that his employer was at fault under two headings:

(1)   common law negligence and
(2)   breach of statutory duty.

Common law negligence has been created and developed by the
judges who have ruled in a series of precedents that an employer
must take reasonable care to prevent his employee, or any other
person within the area of foreseeable risk, from suffering injury to
his person. Where Parliament has imposed duties on an employer
in the criminal law, the judges have in the case of some statutes

implied a right in the person for whose benefit the duty was imposed to bring a civil action for damages in addition to the possible criminal prosecution brought by the public authority. Not all safety statutes will give rise to such a civil action. The Health and Safety at Work Act (HSWA) specifically provides (s. 47) that no civil proceedings shall lie for failure to carry out the duties in sections 2–8 of the Act. A civil action may, however, be brought for breach of statutory regulations made under the HSWA, unless they provide to the contrary (as do the Management of Health and Safety at Work Regulations 1992). The older safety statutes, like the Factories Act, have frequently been held to create civil liability. Most of the leading cases on the interpretation of provisions in these Acts are decisions of civil courts.

7.3 What is negligence?

Negligence is a failure to take reasonable care which causes foreseeable damage. In the context of the employment relationship, the employer has a three-fold duty, to provide safe plant and equipment, safe personnel and a safe system of work. What is reasonable depends on the facts of each case. As with the criminal law, the courts may balance the risk against the cost of avoiding it. In *Latimer* v. *AEC* (1953), a factory floor was flooded by exceptionally heavy rainfall. When the water subsided the floor became slippery from oil which usually ran in a channel in the floor. The employers could have closed the factory, but they decided to continue production after spreading sawdust. Latimer slipped and was injured. It was held that on the facts the employer had acted reasonably and was not liable.

If the employer is aware of a special vulnerability in one of his employees, his duty of care rises. In one case in 1951 (*Paris* v. *Stepney Borough Council*) an employer knew that his employee was blind in one eye. The man was working on a job which involved the slight risk of a chip of metal entering the eye. Other workers were not provided with goggles because the risk was small, but the House of Lords held that the employer was negligent in not giving Paris goggles, because the risk of blindness, though not of accident, in his case was greater (he was blinded when a piece of metal entered his good eye). On the other hand, an employer cannot be reasonably expected to take care if he is not aware of his employee's disability, as in the case of the worker who concealed his epilepsy from his employer, had been warned by his doctor not to work at a height, but fell to his death when he had a fit while working 20 feet above ground; it was held that the worker was 50 per cent to blame for his own death even though the employer was also at fault in not providing full safety equipment (*Cork* v. *Kirby Maclean* (1952)).

There are special rules about defective tools and equipment. Where an employer provides defective equipment to his employee for use in the course of the employer's business and the employee is injured or killed, he or his dependants can sue the employer whether or not the employer has been negligent, as long as there has been fault on the part of a third party, usually the manufacturer of the equipment. The fault of the third party is attributed to the employer for policy reasons: the employee would find it difficult to pursue the manufacturer, often in a foreign jurisdiction, so Parliament has placed that burden on the employer who is better able to bear it. Of course, the employee may also choose, if he wishes, to sue the manufacturer, and the imposition on the manufacturer or importer of strict liability for goods by the Consumer Protection Act 1987, discussed later in this chapter, has made this easier. A recent example of the operation of the Employers Liability (Defective Equipment) Act concerned seamen drowned when their ship sank off the coast of Japan. The House of Lords held that a ship was 'equipment' within the Act, so that the shipowners were liable without negligence on their part for defects in it caused by the negligence of the builder (*Coltman* v. *Bibby Tankers* (1987)). The Act includes any plant or machinery, vehicle, aircraft or clothing and any disease or physical or mental illness. But note that the employer is only liable under this legislation to his own employees and for equipment which he himself provides. Liability to the employees of subcontractors or to the self-employed will still depend on proof of negligence on his part.

Many cases are brought for a failure to provide a safe system of work. It is not enough to see that the premises are safe and that safety equipment is available: the employer must take reasonable care to see that workers are properly trained and supervised. When a worker was blinded because he was not wearing goggles, it was not enough to show that goggles were provided: the employer should have issued orders that they were to be used and the foreman should have supervised the workers to see that instructions were obeyed (*Bux* v. *Slough Metals* (1974)). Though the employer is entitled to expect the workers to take care for their own safety to some degree, it is especially important to warn of latent dangers. In *Pape* v. *Cumbria County Council* (1991) the employee was a cleaner who contracted severe dermatitis through contact with household chemicals. She was provided with gloves, but it had not been explained to her why it was important to wear them. She was awarded £22,000 damages:

> 'The danger of dermatitis or acute eczema from the sustained exposure of unprotected skin to chemical cleansing agents is well known, well known enough to make it the duty of a reasonable employer to appreciate the risks it presents to members of his cleaning staff, but at the same time not so well known as to make it obvious to his staff without any necessity for warning and instruction.'

What can reasonably be expected from the employer depends on a number of different factors. The standard is only that of the average, not of the pioneer. Therefore, it is usually sufficient to follow a practice generally established.

An important test case in the early 1980s raised the issue of the duty of care of employers in noisy industries to take precautions against hearing loss. Albert Thompson was employed as a labourer in a ship-repairing yard for about 40 years. It had been known for over a century that excessive noise caused deafness, but no protection was provided by employers until the early 1970s. By the time Thompson was made redundant in 1983, aged 61, he needed to wear a hearing aid and suffered from a continuous buzzing in the ears. The High Court accepted that his employer was negligent in not providing him with ear protectors at an earlier date, but then had to fix the material date. Suitable devices had been developed by the RAF during the Second World War and were mentioned in medical journals from about 1951, but the first official guidance came with the report of the Wilson Committee in 1963. Thereafter, a Ministry of Labour pamphlet gave advice and in 1972 the Department of Employment issued a Code of Practice. The judge held that an employer cannot be expected to be much in advance of general practice in the industry. 'The employer must keep up to date, but the court must be slow to blame him for not ploughing a lone furrow'. He concluded that 1963 'marked the dividing line between a reasonable (if not consciously adopted) policy of following the same line of inaction as other employers in the trade, and a failure to be sufficiently alert and active ...'. Sadly for Thompson, the medical evidence was that most of his hearing loss had occurred before 1963. He was awarded damages of £1,350 (*Thompson* v. *Smiths Shiprepairers* (1984)).

This decision is but one illustration of the stress placed by judges on official pronouncements. Employers are expected to be aware of statements from the Health and Safety Executive and other official bodies and ignore them at their peril. It is vital to have an efficient method of receiving and taking note of relevant publications. This may be considered a suitable role for the larger occupational health department.

The Northern Ireland Court of Appeal differed from the ruling of the English judge in the Thompson case. They held that even before the publication of *Noise and the Worker* in 1963 there was sufficient medical, scientific and legal knowledge available to shipyard employers to have warned them of the nature and extent of the problem and that some kind of protection was required (*Baxter* v. *Harland and Wolff* (1990)). Northern Irish shipyard workers were awarded damages for hearing loss in respect of exposure back to 1954, the cut-off date fixed by the Limitation Acts (*Arnold* v. *CEGB* (1988)).

Where an employee worked for a succession of negligent employers it is the practice to apportion liability between offending employers (or rather their insurers).

In the last decade there has been a notable rise in the numbers of claims for work-related ill-health, reflecting the increased awareness of the workers both of the hazards and the availability of compensation. The duty to educate and train the employees makes it more likely that they will connect symptoms with their work.

Vibration white finger gives rise to many civil actions. The syndrome is known as Raynaud's Phenomenon after the physician who first described it in 1862. It is caused by exposure to vibration and progresses from intermittent tingling in the hands to extensive blanching, tingling and numbness. The condition has been a prescribed industrial disease since 1981. Hugh Bowman worked in the Northern Ireland shipyards for many years, for sixteen years with vibrating tools (caulking hammers). He often worked in the open, in cold and wet conditions. It was argued that the employer, once he became aware of the risks, should have reduced the hours of work with the hammers, should have provided the workers with specially designed protective gloves and should have given warnings and instituted medical examinations and monitoring, none of which had been done. The Northern Ireland High Court had to determine the date at which a reasonable employer should have become aware of VWF and begun to take precautions. It was held that by 1973 when the condition had been examined in a large number of medical publications, and had been the subject of two reports from the Industrial Injuries Advisory Council, 'a sensible and reasonably well informed safety officer and factory doctor would between them have reached the conclusion that there was something of which they should take note and to which they should alert the management'. Mr Bowman was awarded £2,500 for a Stage 3 condition (on the Taylor scale published in 1982) (*Bowman* v *Harland and Wolff* (1992)). The judge stated that the modern occupational physician should regularly read the *British Journal of Industrial Medicine* and have access to and consult *Hunter's Diseases of Occupation*. Failure to do so might constitute negligence.

In the 1990s work-related upper limb disorders, often called repetitive strain injury (RSI), have achieved prominence. There are two kinds of work which carry this risk: the 'white collar cases' – mainly work with word processors, and the 'blue collar cases' – manual work involving repeated movements of the hand or arm. Despite a statement of Judge Prosser in the High Court in 1993 that the term RSI is 'meaningless' and 'has no place in the medical dictionary', there have been several cases in which employees have been awarded substantial damages against their employers in respect of these conditions, some of which are now prescribed diseases. In 1991 a court awarded two British Telecom computer operators £6,000 each in respect of musculoskeletal injuries sustained while using a keyboard. It was held that in 1981–2 the state of developing knowledge was such that BT could not have been expected to know the full causal link between RSI and keyboard work, but that they should have known of the importance of seating and posture for those engaged in intensive repetitive keyboard work (*McSherry* v. *British Telecom*). The judge dismissed BT's claim that the occupational health department had no means of knowing that there was a problem because no complaint had been made. It was the duty of the department to inform managers of the need to provide proper workstations and seating. In fact, the Chief Medical Officer of BT, who was not called as a witness, was an acknowledged expert on tenosynovitis.

The retiring Chairman of the Health and Safety Commission stated in November 1993 that the Commission had no doubt that RSI exists and can be avoided or minimised by employer action. The courts are likely to agree. In January 1994 a typist was awarded £79,000 against the Inland Revenue in an out-of-court settlement. Her work on an electric typewriter for seven hours a day has left her registered as disabled. The Display Screen Equipment Regulations 1992 will add a possible action for breach of statutory duty to the existing action for common law negligence.

For the future, it is possible that we may see an increase in claims for stress-related disease or injury, especially mental health problems caused or exacerbated by work. Mental illness is as common as heart disease and three times as common as cancer. In 1989/90 80 million working days were lost due to sickness absence certified as mental illness. The courts have held that the employer has a duty to take care to protect his employees against unreasonably high levels of stress and to provide support to those whose work is inherently stressful. A junior doctor alleged that long working hours were damaging his health. The Court of Appeal held that this was capable of being a breach of contract by the employing health authority who had a duty to care for the welfare of its employees. To the extent that the contract provided for excessive hours it was arguably void (*Johnstone* v. *Bloomsbury Health Authority* (1991)). Junior doctor contracts were subsequently varied. And in *Petch* v. *Customs and Excise Commissioners* (1993), it was accepted by the Court of Appeal that employers owe their employees a duty to take reasonable care of both their physical and mental health, subject to the caveat that foreseeability and causation are likely to be more difficult issues in mental injury cases. Though there have been few cases up to now in English courts, the large numbers of such actions in the United States may be the harbinger of things to come.

Another area where there are many claims pending against the employer is damage caused by passive smoking. Since Mrs Veronica Bland, an employee of Stockport MBC, received £15,000 in an out-of-court settlement in 1993, the floodgates have opened in the sense that large numbers of claims are now being pursued. Proof of causation will be a major problem, but less so now that there is more scientific evidence of the dangers of second-hand cigarette smoke, classified by the US Environmental Protection Agency as a first class carcinogen. Several actions are also proceeding against cigarette companies, who are likely to argue that smokers' injuries are self-inflicted.

7.4　Breach of statutory duty

'Old' safety legislation dating from before the Health and Safety at Work Act 1974, like the Factories Acts, invariably provided only criminal sanctions by way of enforcement. Judges in the nineteenth century felt that those injured by acts declared criminal by these statutes should be given a right to compensation, which the common law at that time failed to provide (the tort of negligence only grew to its present stature in the twentieth century). They therefore implied into Parliament's express words an implied right to civil damages for breach of statutory duty. The essence of this tort is that a duty imposed by statute has been infringed, so that only if the case falls within the wording of the statute can the victim sue for compensation. In addition, not all statutory duties give rise to civil actions. Sometimes, the statute expressly confers or denies such an action, but more often it is left to the judges to read the mind of Parliament. With the development of negligence, statutory duty is not now the vital source of compensation that once it was. It is, however, still valuable to plaintiffs, especially in the following respects:

(1) The old statutes spell out the employer's duty in great detail, whereas the common law is expressed in vague terms of reasonableness. It is easier to prove that goggles or a crawling board were not provided than that the employer failed to take reasonable care. This is both a strength and a weakness. Complex technical legislation gives rise to much detailed case law with decisions based on literal interpretation rather than abstract justice. In *Close* v. *Steel Co. of Wales* (1962) it was held that s. 14 of the Factories Act 1961, which at that time imposed a duty on the occupier of a factory to fence every dangerous part of any machinery, only applied to an accident caused by the machine operator being caught in the machine, not by parts of it flying out and striking him. Close also failed to win damages for common law negligence, because the likelihood of a bit on an electric drill shattering and ejecting fragments which flew far enough to damage a worker's eye was so remote that it could not be said that the employer was negligent.

(2) Most statutory obligations demand that the employer or occupier must do that which is reasonably practicable, but sometimes Parliament will impose a duty which is strict or absolute, that is not dependent on proof of negligence. A provision in the Factories Act, 1961, s. 5, that effective provision shall be made for securing and maintaining sufficient and suitable lighting was construed as imposing liability even in a case where a light bulb had unexpectedly failed without fault on the employer's part (*Davies* v. *Massey Ferguson* (1986)). The existence of such obligations in the pre-1974 legislation reflects its basically criminal character. The rationale of imposing criminal liability without fault, as is frequently done in statutes designed to protect the public, is that it induces an organisation to aim at ever higher standards. Proof that a defendant took all possible steps to avoid damage can be taken into account at the stage of imposing sentence after the commission of the crime has been established, by fixing only a nominal penalty. But if the statute is held also to create a civil action for breach of statutory duty, it is not possible to reduce damages because a defendant was not at fault. Judges, unwilling to order a careful defendant to pay large sums in damages, have used their power of interpretation to avoid strict liability, as in the Close case above. The Health and Safety at Work Act contains no offences of strict liability in the parent Act, which in any event expressly excludes any civil action for breach of statutory duty in respect of sections 2–8, though regulations made under the Act to replace existing provisions may impose strict liability.

In Chapter 5 there is a detailed account of the 1992 Regulations made under the HSWA in order to implement the European Framework Directive and the six daughter directives. This legislation departs from the language of reasonable practicability with which English lawyers are familiar. It uses expressions like 'efficient', 'adequate', 'appropriate'. It remains to be seen how English courts will interpret these provisions, but it is likely that they will be drawn to interpretations which reflect ideas of reasonableness with which the English judges are so familiar, rather than creating new categories of strict liability.

(3) The existence of a statutory duty reinforces an allegation of negligence (conversely the absence of such a duty helps to negative an assertion of negligence). 'The reasonable employer is entitled to assume *prima facie* that the dangers which occur to a reasonable man have occurred to Parliament' (Somervell LJ in *England* v. *NCB* (1953)). However, it is possible for an employer to be liable under the statute but not at common law (as in cases of strict liability), and liable at common law but not under the statute (where the words of the statute do not apply to this particular case, as, for example, where premises are held not to fall within the definition of a factory, but there is a failure to take reasonable care). In *Chipchase* v. *British Titan Products* (1956), statutory regulations obliged the employer to provide employees working six and a half feet or more above the ground with working platforms at least 34 inches wide. Chipchase was working at six feet and fell off a platform only nine inches wide. There could be no breach of the statute for that must be literally interpreted, but it could still be held that the employer had failed to take reasonable care and was negligent at common law.

7.5   Liability for the negligence of others

In the civil law, an employer is liable for the negligence of all his employees when acting in the course of their employment. This principle of vicarious liability is founded on the idea that the employer profits from the activities of his employees and in justice should pay for any harm perpetrated by them. An employer may be held liable even though he took every possible precaution. Employers have had to pay damages for the carelessness of an employee contrary to the employer's specific instruction, and even for criminal acts. Only if the employee was 'on a frolic of his own' will he be exempt. As with industrial injuries benefit, the employee is not normally acting in the course of his employment when on his way to or from work unless travel is part of the job. However, in recent case law, the courts have shown increasingly flexibility. In *Smith* v. *Stages* (1989), two workers were employed at a power station in Burton-on-Trent; they were asked to go to Pembroke for a week to work there. The employer paid their expenses and *wages* for the period while they were travelling. While driving back from Wales to his home in Burton in his own car, one employee drove negligently, causing the other's death. The House of Lords held that the employer was vicariously liable.

Disobedience to instructions does not necessarily take the employee out of the scope of his employment. In *Century Insurance Co.* v. *N. Ireland Road Transport Board* (1942), an employee delivering petrol from a tanker lit a cigarette and caused a conflagration. He knew that smoking was strictly prohibited by the employer. Nonetheless, the employer was held vicariously responsible.

Several cases deal with accidents caused by a practical joker. In general, an employer will not be responsible for dangerous tricks perpetrated by an employee, because that cannot be said to be part of the job. In a case in 1987, a young woman was in the washroom at work when a fellow employee as a joke rocked one of the wash basins which was known to be slightly unstable in her direction. The plaintiff turned quickly and strained her back. The Court of Appeal applied the following test:

'... a master is responsible not merely for what he authorises his servant to do, but also for the way in which he does it. If a servant does negligently that which he was authorised to do carefully, or if he does fraudulently that which he was authorised to do honestly, or if he does mistakenly that which he was authorised to do correctly, his master will answer for that negligence, fraud or mistake.'

It was held that, although visiting the washroom was within the scope of employment, pushing the wash basin was outside employment and thus the employer was not liable (*Aldred* v. *Nacanco* (1987)).

On the other hand, the employer will always be liable for his own negligence. When he knows that a particular employee is a danger to others his duty of care dictates that he should warn him, or in a bad case dismiss him. And in the case of the washbasin, the employer would have been liable if there had been evidence that its instability made it hazardous if properly used.

Before 1948, English law applied a principle known as the doctrine of common employment. This held that no employee could hold his employer vicariously liable for an act of negligence of a fellow worker in the same employment. Judges tried to avoid this doctrine by stressing the primary duties of the employer which, they held, could not be delegated to anyone else. If an employee was injured by the negligence of someone to whom the employer had delegated the performance of one of his primary duties, the employer would still be liable, for this would be a breach of a duty placed directly on the employer, not an example of vicarious liability. The leading case (*Wilson's and Clyde Coal Co.* v. *English* (1938)) concerned an employer who entrusted the task of providing a safe system of work to a reliable and competent employee. Because of an uncharacteristic act of carelessness on the latter's part, a fellow employee was injured. The employer could not be held vicariously liable because of common employment, nor was he negligent himself because care had been taken in selecting a delegate. The House of Lords held the employer liable for breach of a *personal* duty to see that care was taken by the person whom he appointed.

The abolition of common employment by the Law Reform (Personal Injuries) Act in 1948 pushed this idea of non-delegable duty into the background, whence it has been resurrected in cases involving a different problem: liability for the negligence of contractors. Vicarious liability in general makes the employer responsible for employees, but not for independent contractors. It would seem that, at least in some situations, the employer can be held responsible to his employee for the negligence of a contractor, as in *McDermid* v. *Nash Dredging* (1987), where an employer 'lent' his employee to another employer who negligently failed to provide a safe system of work. Lord Brandon said this:

> 'The essential characteristic [of the employer's duty of care] is that, if it is not performed, it is no defence for the employer to show that he delegated its performance to a person, whether his servant or not his servant, whom he reasonably believed to be competent to perform it.'

For example, where an employer has delegated to an occupational health department the performance of his duty of care towards his employees, like the duty to provide appropriate health surveillance, it probably makes no difference whether he employs his own in-house staff, or buys in the services of an independent contractor. If the doctor or nurse is negligent, the employer will be vicariously liable.

This principle will apply as a general rule only where the contractor works in close conjunction with the defendant employer and it does not make the employer automatically responsible for a contractor's defective premises. In *Cook* v. *Square D. Ltd* (1992) the employee was an electronics engineer who was sent on assignment to Saudi Arabia to work on a contractor's site. There, he tripped on an unguarded hole in the flooring and sustained injury. It was held that his employer was not liable for the state of contractors' premises over which it had no control. However, one judge was of the opinion that where a number of employees are going to work on a foreign site, or where one or two employees are called on to work there for a very considerable period, an employer may be required to inspect the site and satisfy himself that the occupiers are conscious of their obligations concerning the safety of people working there.

> 'The master's own premises are under his control: if they are dangerously in need of repair he can and must rectify the fault at once if he is to escape the censure of negligence. But if a master sends his plumber to mend a leak in a respectable private house, no one could hold him negligent for not visiting the house himself to see if the carpet in the hall creates a trap. Between these extremes are countless possible examples in which the court may have to decide the question of fact: did the master take reasonable care so to carry out his operations as not to subject those employed by him to unnecessary risk? (Pearce LJ in *Wilson* v. *Tyneside Window Cleaning Co.* (1958)).

7.6 Causation

The plaintiff must prove negligence and/or breach of statutory duty and he must show that he suffered damage as a result. He must prove that, on a balance of reasonable probability ('more likely than not'), his injury was caused by the defendant's fault. In *McGhee* v. *NCB* (1972), a labourer cleaned out brick kilns. Although he was exposed to clouds of brick dust, the employers did not provide showers and McGhee had to cycle home daily caked with sweat and grime. He contracted dermatitis after five days in these conditions. The court found that the employers were not negligent in requiring him to work in dusty conditions, but failed in their duty in not supplying adequate washing facilities. The problem was that the doctors could not state with confidence that he would certainly not have contracted dermatitis if the employers' plumbing had been first class, though willing to give evidence that the failure to provide showers materially increased the chance that dermatitis might set in.

In this case the employers were held liable, but the decision has been questioned in a subsequent action for medical negligence outside the field of health and safety at work (*Wilsher* v. *Essex AHA* (1988)). A premature baby was given too much oxygen soon after his birth, allegedly because of the negligence of his doctors. He was eventually found to be suffering from retrolental fibroplasia, a condition which might or might not have been caused by the excess oxygen. Other possibilities were hypercarbia, intraventricular haemorrhage, apnoea or patent ductus arteriosus. The House of Lords held that is for the plaintiff to satisfy the court that the negligence of the defendant materially contributed to the plaintiff's injury (even if not the sole cause). The judge must not assume that the labourer's dermatitis or the child's blindness was caused by the defendant's negligence, but must ask for evidence that this was a likely cause. The House of Lords was satisfied that such evidence existed in the McGhee case, but sent the Wilsher case back for a retrial.

Similar issues arose in *Bryce* v. *Swan Hunter* (1988). Bryce had been employed by the defendants for only part of his working life. It was established that during these periods the defendants had negligently and in breach of statutory duty exposed him to excessive amounts of asbestos dust. Bryce contracted mesothelioma, a cancer caused by exposure to asbestos dust and proved to be dose-related, though the medical profession does not know precisely why an increase in exposure adds to the risk of contracting the disease which is caused by one fibre entering the pleura. It may be that increased exposure merely increases the statistical risk. Did the fatal fibre which caused the cancer arise from the defendants' negligence, or was it a fibre for which they were not responsible, one of those inhaled when he was working for another employer or within the acceptable level of exposure at the time? It was, of course, impossible ever to identify the source of the rogue fibre and if the law demanded that degree of proof the plaintiff would be bound to fail. The judge held that it was sufficient to show that the defendants' negligence materially increased the risk that the plaintiff would contract mesothelioma, so the plaintiff won his action.

Epidemiological evidence may be important. In *McPherson* v. *Alexander Stephen and Sons Ltd* (1990) the plaintiff was exposed to excessive levels of asbestos dust through his employer's negligence. He neither smoked nor drank heavily. He contracted cancer of the larynx. The Scottish court heard evidence that there was an increased occupational risk of 1.3 to 2.0 on a 95 per cent confidence basis of cancer of the larynx from asbestos. The court was satisfied that the plaintiff had established causation.

The law of tort in most cases imposes liability only for damage which the defendant could foresee. If a reasonable man could not have appreciated that his acts or omissions would cause harm, he cannot be held liable for his failure to take precautions. Thus, a failure by a doctor to appreciate that ampoules in which a spinal anaesthetic was stored in a solution of phenol had developed cracks so fine that they could not be seen or felt, but which allowed the phenol to contaminate the drug, was not tortious because this danger was unknown to the medical profession at that time (*Roe* v. *Minister of Health* (1954)).

If, however, the defendant was negligent, he is liable for all the damage which results, however unforeseeable in its extent. The best known example of this principle is the maxim: 'The wrongdoer must take his victim as he finds him'. In *Smith* v. *Leech Brain Co.* (1962), a workman while working with molten metal was burned on his lip from a fleck of metal splashed on to it. His employer was negligent in not providing a shield. The wound later became malignant and the employee died of cancer, a result totally unforeseen at the time of the accident. The employer was liable for the death, because his negligence had caused the burn: his ignorance of the employee's predisposition to cancer was irrelevant. In *Bradford* v. *Robinson Rentals* (1967), a lorry driver contracted frostbite when he had to drive his employer's unheated lorry in bad weather for a long distance. Despite the unusual severity of his reaction to an English winter, he was able to recover full compensation from his employer. The principle also applies when the physical extent of damage was unforeseeable, so that a defendant will be responsible for a catastrophic fire if he could have foreseen a small one.

Negligence in itself does not give rise to civil liability: damage must result. Thus, a doctor who negligently gave an injured employee the full dose of ATS, an anti-tetanus injection, without waiting for half an hour to test his sensitivity, was not liable for his resulting encephalitis, since that developed more than a week after the injection. The employer whose carelessness had caused the original wound was held liable for all the damage because he had to take his victim as he found him, with a special susceptibility to the vaccine (*Robinson* v. *Post Office* (1974)). If the encephalitis had developed soon after the injection, the doctor would have been liable for the encephalitis and the employer only for the wound, because the doctor's negligence would in effect have broken the chain of causation, being what lawyers term a *novus actus interveniens* (a new intervening event).

Courts occasionally assume that there has been negligence if an injury has been suffered in circumstances where there should have been no risk. The patient who leaves the operating theatre with a surgical instrument in her abdomen will argue that there must have been negligence: the thing speaks for itself (*res ipsa loquitur*).

7.7   Liability of the employer to non-employees

The law of negligence imposes a duty of care in many situations towards anyone within the foreseeable area of risk. A factory owner has to consider his own employees, employees of contractors, visitors to the premises and members of the public both on his premises and within the vicinity. In *Haley* v. *London Electricity Board* (1965), the Board was held liable to an unaccompanied blind man who fell into a hole in a public pavement because it was insufficiently guarded. The House of Lords held that the Board's employees should have foreseen that blind persons were in the community and taken special precautions for their benefit.

The duty of care of an occupier of land in respect of premises has been enacted in the Occupiers Liability Acts 1957 and 1984. The occupier may be the owner of the premises or a contractor in control of a site. The first statute deals with liability to *visitors*, those with express or implied permission to be on the premises, the second with liability to *trespassers*, those who enter the premises without the permission of the occupier. A common duty of care is owed to all visitors, a duty to take reasonable care in all the circumstances of the case. Relevant considerations are that an occupier must be prepared for children to be less careful than adults and that he is entitled to expect that someone who enters his premises to do a job will appreciate and guard against the usual risks of the job. A warning of a danger does not exonerate an occupier unless it is sufficient to enable the visitor to be reasonably safe and the Unfair Contract Terms Act 1977 makes ineffective a notice excluding liability in respect of business premises.

The occupier is not liable for the negligence of an independent contractor, but he will be responsible for his own negligence in choosing a contractor who he should have known was unreliable, or failing to exercise reasonable supervision. In the case of *Ferguson* v. *Welsh* (1988), Ferguson was injured when working on a site occupied by a local authority. He was employed by a demolition subcontractor who had been brought in by the main contractor and who had adopted a thoroughly unsafe system of demolishing a building from the bottom up. It was held that the council were not liable under the Occupiers' Liability Act. They were unaware that the main contractor had employed a 'cowboy' contractor (it had been done without their permission) and it is not normally reasonable to expect an occupier of premises to supervise the contractor's activities unless he has reason to suspect that the contractor is adopting an unsafe system of work.

The 1984 Act imposes a duty towards trespassers, but an occupier is only liable if he knows, or has reasonable grounds to believe, that a danger exists and that trespassers may come into its vicinity. The plaintiff must also prove that the risk was one against which the occupier can reasonably be expected to provide protection. The Act is designed to give statutory force to cases like *British Rail* v. *Herrington* (1972), where damages were awarded to a small child who was injured on a live rail to which he had gained access through a gap in the defendants' fence of which they knew.

7.8   Duty to the unborn child

First, it must be remembered that a woman with child is, in effect, two people, both of whom may wish to seek compensation. The tragedy of the thalidomide children made the community conscious of the vulnerability of the child in its mother's womb. At the time, lawyers were uncertain whether an action could be brought by a child for damage sustained before birth. Parliament clarified the law in the Congenital Disabilities (Civil Liability) Act 1976 and since then the court in *Burton* v. *Islington Health Authority* (1993) has held that children born before 1976 have a cause of action at common law. An action only lies under the Act if the child is injured by an act or omission which is also actionable by one or other of the child's parents. The child must be born alive. A foetus killed in the womb has no rights under this Act, but the *mother* has a right of action at common law if she miscarries because of another's negligence, as where she can prove that exposure to a substance or process has caused a spontaneous abortion. Damage to the reproductive capacity of a male worker through radiation or toxic chemicals, if it can be proved, may also give rise to an action for damages.

The legislation is not limited to an injury to a child after conception, but extends to an occurrence which affects either parent in his or her ability to bear a normal healthy infant and which leads to the birth of a disabled child, *unless at the time of conception either parent knew of the risk of disablement.* Again, the parents can also sue at common law for negligence which affects their ability to have children. The mother cannot be legally liable to her own baby if her conduct (drinking, smoking, taking drugs etc.) causes its disability, except that a woman driving a motor vehicle when she knows or ought to know herself to be pregnant is under a duty of care towards her unborn child. The father has no such immunity.

In the employment situation, therefore, the employer will have to consider the health of father, mother and child. If, through his negligence, a woman worker loses her baby or is advised to have an abortion, the mother will be able to claim damages. If his negligence causes a deformity in the child of either a male or female worker, both child and parent will be able to sue, the child for his disability and the parent for psychological distress. English courts have refused to give damages for 'wrongful life' to a child whose complaint was that her mother's doctor was negligent in not informing her that the foetus was defective so that she could have aborted her (*McKay* v. *Essex AHA* (1982)).

The 1976 Act includes a special section for health professionals:

'The defendant is not answerable to the child, for anything he did or omitted to do when responsible in a professional capacity for treating or advising the parent, if he took reasonable care having due regard to the then received professional opinion applicable to the particular class of case; but this does not mean that he is answerable only because he departed from received opinion.'

This sounds very like the standard of reasonable care discussed in Chapter 2.

A report by Professor Martin Gardner in 1990 explaining the higher than average incidence of childhood leukaemia and Hodgkin's disease in children born to fathers who had worked at the Sellafield nuclear processing plant as caused by pre-paternal irradiation was relied upon by two plaintiffs in an action against British Nuclear Fuels in 1993 (*Hope and Reay* v. *BNFL*). The High Court rejected the claim, finding insufficient proof of causation on the facts. Shortly after, the HSE published a report of an investigation into leukaemia and other cancers in the children of male workers at Sellafield (1993) which concluded that weak evidence had been found for the Gardner conclusions.

7.9   Post traumatic stress disorder

No damages will be paid for mental suffering alone, but if as a result of fear or terror the plaintiff becomes ill, compensation can be given for that illness. It must be shown that the defendant's negligence caused a shock which was reasonably foreseeable in an average phlegmatic individual. Doctors have developed better understanding of post traumatic stress disorder, and this has been reflected in an increase in the numbers of claims under this head.

The law on nervous shock was reviewed by the House of Lords in *Alcock* v. *Chief Constable of the South Yorkshire Police* (1991), one of the cases arising out of the disaster at the Hillsborough football stadium. Due to the negligence of the police in controlling the crowd too many spectators were admitted to the enclosure and some were crushed to death. The events were broadcast on radio and television as they occurred and were seen and heard by many relatives and friends of the deceased. Others were eye-witnesses at the ground or identified bodies in the mortuary.

The House of Lords decided that the mere fact that nervous shock to the plaintiff was reasonably foreseeable did not give rise to a duty of care, because of the multitudinous claims which might result. The law had to limit the remedy to those 'proximate' to the accident, either close relatives present at the ground or those directly involved in the events, like rescuers. Those who saw the incident on television or heard it on radio could not be included, nor those who saw the victims some hours later and not in the immediate aftermath.

This ruling does not interfere with the authority of prior cases like that of a crane driver who, without any fault on his part, injured a fellow worker while operating a defective crane and suffered a psychiatric illness. He recovered damages from those responsible for the defect (*Carlin* v. *Helical Bar* (1970)). Damage to property may give rise to nervous shock as in the case of the lady whose house burned down because of the negligence of British Gas and who recovered damages for the nervous illness caused by the spectacle of the conflagration (*Attia* v. *British Gas* (1987)). Both these plaintiffs were more than spectators: they were themselves directly involved. Compare these with *McFarlane* v. *E. E. Caledonia* (1993) where a painter on a support vessel suffered psychiatric illness after he witnessed the explosion and fire on the *Piper Alpha* oil rig, though not in danger himself. He lost his action for compensation.

Where the plaintiff suffers a physical injury due to the defendant's negligence and in consequence suffers a psychiatric illness, both physical and mental effects are compensable. In *Pigney* v. *Pointer's Services* (1957), the negligence of the defendants caused head injuries to the plaintiff's husband, which induced depression. He eventually committed suicide and it was held that the defendants were liable for his death.

In *Sykes* v. *Ministry of Defence* (1984), the plaintiff had been employed in the dockyards all his working life. Due to his employer's negligence, he had been exposed to excessive amounts of asbestos dust. He was told by his doctors that his X-rays showed calcified pleural plaques, but no asbestosis or cancer. The plaques gave rise to no symptoms, but were evidence of his exposure to asbestos. The judge gave damages of £1,500 for the physiological damage and the plaintiff's anxiety that he might develop more serious illness at a later date.

7.10 Assumption of risk and the fault of the plaintiff

Tort liability being dependent on fault, the carelessness of the plaintiff for his own safety will sometimes reduce, sometimes preclude damages. As we have seen, it will not usually deny him no-fault industrial injury benefit unless his actions take him out of the scope of his employment. *Volenti non fit injuria* is a Latin maxim which means that no-one can sue if a risk, willingly assumed, materialises. A participant in a sport cannot sue the player who tackles him according to the rules of a game, just as a patient cannot sue a doctor for performing an operation to which the patient consented. Yet neither the player nor the patient has assumed the risk of play or treatment outside the rules and thus could sue for an unlawful tackle or a negligent performance of the operation.

Rarely does the law deny a remedy to an injured worker on the basis that he has voluntarily assumed the risk of injury, because it appreciates that one in an inferior economic position is not free to choose. The labourer who worked under a crane loaded with stones was able to recover damages when a stone fell on him: his knowledge of the risk did not mean that he had accepted it (*Smith* v. *Baker* (1891)). A health professional is deemed to have accepted the risk that he may contract an infection through contact with patients, but not if this is the fault of the employer who has been negligent in not supplying proper equipment, sufficient qualified staff or a safe system of work. If the risk is significant, as with some workers in potential contact with hepatitis B, the employer will be negligent in not providing vaccination, if available. Occasionally, when a risk is relatively minor and the plaintiff accepts the job with full knowledge of it, the employer will not be negligent, as in *White* v. *Precision Castings* (1985) in which there was evidence that it was well known to grinders that they would suffer from Raynaud's phenomenon, a numbness of the fingers causing minor discomfort and minor inconvenience. In *Withers* v. *Perry Chain* (1961) an employee after five years in the same job had to stop work because of dermatitis. The employers had taken all possible precautions, but the employee developed an allergy. There was no over-exposure to any substance, merely the development of an abnormal sensitivity in the employee. The employers moved her to a different job, but the dermatitis returned. She sued the employer for permitting her to continue in the second job. The Court of Appeal held that there was no duty to dismiss or refuse to employ an adult employee who wished to do a job because of a slight risk of which the employee was fully aware.

The defence of assumption of risk cannot apply where the employer is in breach of his statutory duty unless the employer has done everything within his power to comply with the statute, when he may escape responsibility if an injury is caused solely by the employee's failure to take precautions in breach of the employee's own statutory duty. In *ICI* v. *Shatwell* (1965), the brothers Shatwell, both of whom were trained and experienced shotfirers, disobeyed all their instructions in carrying out an electrical circuit test in the open without taking cover. They were both injured. The employers had provided safety equipment and safety training: there was nothing more that could have been done by them. The employees' conduct constituted a breach of statutory duty placed on them by regulations made under the Mines and Quarries Act. It was held that neither brother could recover damages. Where an accident is caused solely by the carelessness of the injured employee himself and the employer is not at fault courts are unwilling to impose civil liability.

Far more common is the reduction of the amount of the plaintiff's damages for contributory negligence (Law Reform (Contributory Negligence) Act 1945). Where any person suffers damage as a result partly of his own fault and partly of the fault of any other person, his damages are reduced to such extent as the court thinks just and equitable. In *O'Connell* v. *Jackson* (1972), the defendant was wholly to blame for colliding with the plaintiff on his moped, but paid only 85 per cent of the damages, because the plaintiff's failure to wear a helmet had made his injury more severe. The plaintiff is guilty of contributory negligence if he has failed to exercise reasonable prudence. One employee who contracted dermatitis because he failed to use the protective cream provided by his employer was held only 5 per cent to blame, because the employer had not insisted on its use (*Clifford* v. *Challen* (1951)). A foundry worker who had been provided with safety goggles, but not instructed that he must wear them nor reprimanded for not wearing them, was held 40 per cent to blame when his eyes were damaged by molten metal (*Bux* v. *Slough Metals* (1974)). A miner who was told that he must bring down an unsafe roof before working under it and disobeyed instructions so that he was killed when the roof collapsed on to him was held 80 per cent responsible (*Stapley* v. *Gypsum Mines* (1953)). In such cases, the court assesses the full amount of the damages and then reduces them by the appropriate percentage.

7.11 Several potential defendants

Very often a plaintiff will have a possible right to compensation from more than one defendant. A worker injured in an accident caused partly by the negligence of his employer and partly by that of a third party may sue either or both. Each defendant has the right to a contribution from the other: the amount of contribution is such proportion as is just and equitable, having regard to the extent of each party's responsibility for the damage in question (Civil Liability (Contribution) Act 1978). The plaintiff is, however, entitled to his full damages, so that if one defendant has no money the other will have to pay the full amount.

Until 1990 it was the practice for health authorities sued as employers of negligent doctors to recover all or part of the damages by way of contribution from the doctors' defence organisations. This practice has now been abandoned under a system known as 'Crown indemnity', whereby the health authority takes full responsibility for defending the action and paying any compensation agreed or awarded.

7.12 Attempts to exclude liability

In these days when actions for negligence are increasing, potential defendants try to protect themselves with notices and contract clauses excluding liability. 'Visitors enter these premises at their own risk.' 'The company does not accept responsibility for the employees of subcontractors.' 'Employees should take note that the employer accepts no liability for loss of or damage to their personal effects brought on to the employer's premises.'

English law in general does not permit anyone to exclude liability for negligence causing death or personal injury, except in his private and family relationships. The Law Reform (Personal Injuries) Act 1948 renders void any provision in a contract excluding or limiting the liability of an employer for personal injuries caused to an employee or apprentice by the negligence of persons in common employment with him. The Unfair Contract Terms Act 1977 makes void any contract term, or any notice given to persons generally or to particular persons, which attempts to exclude or restrict liability for death or personal injury resulting from negligence committed in the course of a business or arising from the occupation of premises used for business purposes. Business includes a profession and the activities of any government department or local or public authority. In *Johnstone* v. *Bloomsbury H.A.* (*supra*) it was said that a contract with a junior doctor was void to the extent that it required him to work hours so long that his health was damaged. The Act also provides that, in the case of other loss or damage, namely to property, a person cannot so exclude or restrict his liability for negligence except in so far as the term or notice satisfies the requirement of reasonableness. Note that the statute is inappropriately named, for it extends to attempts to exclude liability by notice where there may be no contractual relationship, like a notice in an NHS doctor's surgery, or a DSS office.

What is a reasonable exemption clause in a case of damage to property is left to the courts, guided by s. 11 of the Act, which states that it is for those alleging that a provision is reasonable to prove it.

7.13   Product liability

We have seen that the Health and Safety at Work Act imposes criminal law duties on designers, manufacturers, importers and suppliers of goods. We now turn to the liability of these persons in the civil law. The *fons et origo* of the law of negligence was the decision of the House of Lords in *Donoghue* v. *Stevenson* (1932). In that case, it was held that a manufacturer of goods owed a duty of care not just to the person who purchased the goods directly from him, with whom he had a contract, but also to the ultimate consumer of those goods in the law of tort. However, there was a vitally important difference between the duty in tort and that in contract. In tort, the consumer must prove that the manufacturer failed to take reasonable care and thereby caused foreseeable damage, but where there is a contract for the sale of goods, the Sale of Goods Act 1979 imposes a duty on the seller of goods in the course of a business to provide goods of merchantable quality, a duty which is strict in the sense that it is not merely to take reasonable care. If I buy a bottle of aspirin tablets from my local pharmacist and suffer damage because the tablets have been carelessly manufactured, I can recover damages from the pharmacist whether or not I am able to prove that he should have known of the defect. We have a contract and he is strictly liable. He may then in his turn be able to sue his supplier under a similar principle, so that the manufacturer may end up paying the bill. But if I want to sue the manufacturer, with whom I have no contract, directly, English law until very recently demanded that I prove negligence, a hurdle at which many claimants, like the thalidomide children and the Opren claimants, have stumbled over the years. In addition, if drugs are dispensed under the NHS, there is no contract with the pharmacist and negligence must be proved to render him liable, and only the purchaser of the goods can sue in contract: his family and other third parties must sue in tort.

The problems of consumers in attempting to obtain compensation led American judges to create a form of strict liability imposed on manufacturers with regard to the ultimate purchasers of their goods. This came to be known as product liability. The

European Community took up this idea and agreed that it be introduced into the laws of all the Member States. The Consumer Protection Act 1987 is the UK's response to the EC Product Liability Directive and came into force in March 1988. The Act does not destroy the old rules of sale of goods and negligence but creates additional remedies for consumers. It is likely to be inflationary in effect in the sense that manufacturers will increase their insurance cover (though insurance is not compulsory) and pass on the costs of the higher premiums by raising their prices.

The Act only applies to goods and electricity. It has no effect on the law relating to the provision of services, although the European Community has proposed the introduction of legislation similar to that on product liability. 'Goods' includes substances (any natural or artificial solid, vaporous or liquid substance) and ships, aircraft and vehicles. It is possible that it also includes blood and human blood products. It does not extend to land or buildings, though products used in the erection of buildings like bricks or cement are covered. The UK has also taken the option provided in the Directive of excluding primary agricultural products and game. Strict liability is imposed in respect of industrially processed food, like tinned meat, but not for the fresh joint of beef on the butcher's slab. Injury or damage arising from nuclear accident is excluded: in the UK this is already covered by the Nuclear Installations Act 1965.

Liability is imposed where any damage is caused wholly or partly by a defect in a product. Anyone injured by defective goods may claim. If a vehicle is sold with defective brakes and causes an accident in which the owner, his passenger and a passer-by are injured all three can sue the manufacturer. Damage means death or personal injury or any loss or damage to property (including land), but only if that property is both of a kind ordinarily intended for private use, occupation or consumption and was in fact intended by the person suffering the damage mainly for his own private use. The Act is designed to protect the private consumer against personal loss, not the business against commercial loss. Very small losses are excluded. If the plaintiff is suing for damage to property under the Act, he must show that his total damages (including any for personal injury) would exceed £275. But remember that anyone injured by defective goods still has the option of suing for negligence or a breach of the Sale of Goods Act, if applicable.

Not all injury caused by goods attracts strict liability. Knives cut fingers as well as carrots, people fall off safe ladders, whisky kills alcoholics, but it does not follow that those products are necessarily defective. The consumer will no longer have to prove negligence, but the burden will be on him to show a defect in the goods. According to the Act, there is a defect in goods if the safety of the product is not such as the public is entitled to expect, taking into account the way the product has been presented and marketed, including any instructions or warnings attached to the goods. The courts should also consider whether the goods have been used in a manner which could reasonably be anticipated (the manufacturer of surgical scissors would not expect them to be given to small children for cutting out pictures) and the time at which the product was first put onto the market (a car without rear safety belts is not defective if, at the time it was first supplied, they were not mandatory). The introduction of the legislation will be likely to lead manufacturers to attach more warnings to their goods. Cigarettes and plastic bags are already labelled: perhaps we shall soon see bottles of whisky with the message: 'This product makes you drunk if you drink too much of it!'

Who is liable for defective goods? The Consumer Protection Act identifies four classes of potential defendant:

(1)   the producer;
(2)   the 'own-brander' who, by attaching his name or trade mark to goods, holds himself out as the producer;
(3)   the importer who brings goods into the EC and supplies them in the course of a business;
(4)   the supplier.

Producers include manufacturers, processors of natural products, and suppliers of component parts and raw materials incorporated in manufactured or industrially processed products. Where Manufacturer A produces a defective tyre which is then built into a car by Manufacturer B, the tyre and the car are both defective, so that, if the car causes an accident, both A and B would be liable. No doubt, the contract between them would provide for an apportionment of liability in such a case. If the supplier of the component or raw material can show that the defect in the finished product was due solely to the design or specifications of the producer of the finished product, he will have a defence. Suppliers (often retailers) are only liable if they cannot identify to the consumer the producer or importer of the goods. If they can do this, their liability under the Act ceases, though they may still have obligations under the Sale of Goods Act.

It is important to note that the strict liability imposed by this legislation only attaches to goods after they are put on the market, unlike in the United States. Therefore, if an employee of a manufacturer is injured by goods in the course of their manufacture or testing by his own employer, he will have to prove negligence in order to obtain compensation. But if an employee is injured by a defect in tools, equipment, vehicles, food and so on supplied to him or his employer by another manufacturer or importer, he will be able to sue that supplier under the Consumer Protection Act.

The Act provides a number of possible defences. One which still causes considerable debate is the development risks defence. The producer of defective goods will be strictly liable for damage caused by those goods *unless he can prove that the state of scientific and technical knowledge at the time the goods were supplied was not such that a producer of products of that kind might be expected to have discovered the defect.* The manufacturers of thalidomide would have had a defence under this provision if they could have proved that medical science was not aware of the risk to the foetus at the time the drug was marketed. The UK has included this defence in its legislation (it is optional and other Member States have excluded it) because of pressure from the insurance industry, fearful of open-ended liability especially in the pharmaceutical industry. The EC Commission protests that the wording of the 1987 Act is too widely drawn and is in breach of the Directive because it alludes to the knowledge of a reasonable producer rather than a reasonable research scientist. An action is proceeding to the European Court.

Critics protest that British consumers have been shabbily treated. A drugs company with a new drug to try may choose to sell first in the UK to take the benefit of the defence. Note that, even with the defence, the consumer is in a better position than before the Act, in that the burden of proof has been shifted to the defendant.

Other possible defences are that the producer was complying with a statutory requirement, that the defect did not exist in the goods when they were supplied but was caused by the way they were subsequently treated, and contributory negligence. An action must normally be commenced within three years of the damage complained of, unless the delay was caused by the consumer's ignorance of material facts. In any case, no action can be brought more than ten years after the date when the product was first put into circulation. Potential defendants are precluded from limiting or excluding their liability by contract or notice.

7.14   Damages

In the tort system, damages must be given as a lump sum payment calculated to provide, when properly invested, for the relevant period, and taking it into account that tax will have to be paid on the income. Our courts do not, however, make any attempt to predict future inflation. When the victim survives, he may need income support for the rest of his life, and when he has been killed, his surviving family who were formerly dependent on him will need another source of income. The court has no power to enter judgment in the form of periodical payments or to order the payment of an annuity, as in the social security system.

The parties, on the other hand, may agree a settlement whereby the plaintiff receives an annuity backed by insurance. The Inland Revenue has agreed to treat these annuity payments as tax-free payments of capital. 'Structured settlements' as they are familiarly known are fast increasing in popularity and have been approved by judges in several cases. The principal advantage for the plaintiff is that he will be provided for throughout his life, for the defendant that he usually ends up paying less. But at present both parties have to agree, otherwise the judge must award a lump sum.

It is now possible for courts to make a provisional award of damages, allowing the plaintiff to return for more at a later date if his medical condition deteriorates (Supreme Court Act 1981). The reason for this new power is to try to settle compensation at an earlier date. Suppose an employee is injured in an accident at work. He seems to have fully recovered after a year, but his doctor writes that there is a possibility of epilepsy developing at a later date. Formerly, his lawyers might advise him that he should wait until his medical condition became clearer, but now he can safety accept a settlement approved by the court, reserving the right to come back if his health deteriorates. This option has little effect in most cases, because they are settled out of court by insurance companies who aim for finality in their agreements.

The tort system involves the use of a crystal ball in two respects. First, the judge must guess what would have happened had the damage not occurred. Would the victim have been promoted? How long would he have lived? Secondly, estimates will have to be made of what the future now holds. Will he make a complete recovery? Will he continue to suffer pain? If the victim is now retraining for a different job, how likely is it that he will be successful, and what sort of earnings will he be likely to be able to command? Formerly, judges used to estimate how likely it was that the grieving widow would be able to attract another husband, but this is now forbidden by statute. Even if the widow has remarried the court cannot take this into account in assessing her damages, though it can reduce her children's compensation because they have a new father. The dependent widower has no similar protection.

7.15   The living plaintiff

He is entitled to *restitutio in integrum* (full compensation). In 1979, a senior registrar aged 36, soon probably to be promoted to consultant, who suffered irreparable brain damage while undergoing a minor operation due to the negligence of staff employed by a health authority, was awarded £250,000, though barely sentient and without dependants (*Lim Poh Choo* v. *Camden and Islington HA* (1979)). Damages are awarded for financial loss, like loss of earnings, both past and prospective (including those for the years which he will not now live because of his likely premature death), loss of earning capacity in a competitive job market (*Smith* v. *Manchester Corporation* (1974)), but also for injury less easy to quantify, like pain and suffering and loss of amenity (in the industrial deafness cases this included watching TV, chatting with friends and hearing the birds sing). The victim of negligence is entitled to seek private medical treatment if he wishes; the possibility of using the NHS is to be disregarded (Law Reform (Personal Injuries) Act 1948). On the other hand, the plaintiff is under a duty to mitigate his loss if possible by, for instance, undergoing medical care which would improve his condition or seeking alternative employment. This does not oblige an injured person to submit to an operation which carries some substantial risk. If the plaintiff is or is likely in the future to be maintained at the public expense in a hospital or other institution, that will be taken into account.

In assessing damages for non-pecuniary loss, the courts have to put a money figure on the loss of a hand or a leg, in the same way as in the social security system the medical panels must assess the degree of disablement. It is important for there to be consistency in the level of awards, both because equality of treatment is just and because insurance companies need some kind of tariff to be able to negotiate settlements. The exclusion of juries in personal injury cases means that the judges have been able to work out a reasonably consistent scheme. This is not to ignore the arbitrary nature of undertaking an assessment of the value of pain and suffering or loss of a limb. The level of tort damages is not linked to disability pensions under the industrial injury system. Further, tort damages are tailored to the circumstances of the individual plaintiff. A worker who loses his hand obtains higher damages if he was an enthusiastic amateur artist in his spare time than if he did nothing but work and watch television. In the social security system, as we have seen, the assessment is made objectively, only taking into account age and sex.

The Pearson Commission recommended that plaintiffs should not be compensated twice, by the defendant and by social security payments. The Social Security Administration Act 1992 provides that virtually all benefits payable for disability and unemployment (including statutory sick pay, disablement pensions, income support, unemployment benefit, disability living and working allowances) for five years from the onset of disease or injury shall be deducted from the damages paid to the plaintiff. The Act excludes damages of less than £2,500, damages for fatal accidents, compensation paid by British Coal under the Pneumoconiosis Compensation Scheme, payments in respect of sensorineural hearing loss where the loss is less than 50 dB in one or both ears, and any payment made under the NHS (Injury Benefit) Regulations 1974. The defendant does not gain: he has to pay to the Secretary of State the amount of benefit received by the plaintiff.

Redundancy payments will be deducted from the damages the defendant has to pay only if the plaintiff's injury is the sole cause of his redundancy. In *Colledge* v. *Bass Mitchells and Butlers* (1988), Colledge was injured at work when he and a fellow worker were carrying a pallet. A fork-lift truck skidded towards them, the other man dropped the pallet and the whole weight came on to Colledge who severely injured his back. After the accident, Colledge was offered and accepted voluntary redundancy. He would have been unlikely to have been made redundant and would have worked for the defendants until his retirement if the accident had not occurred. Therefore, it was held that the redundancy payment was a benefit which must be set off against the damages awarded for the employers' negligence.

Public policy considerations cast a different light on the deduction of benefits from private insurance policies. If the employee has taken out his own insurance it is thought wrong to punish his prudence by deducting the insurance money from his damages. Why should his negligent employer benefit from his premiums? The same applies if the plaintiff receives a payment from a charity or a disaster fund or some other charitable third party, like his trade union. The rule also extends to insurance provided by the benevolent employer. In *Parry* v. *Cleaver* (1970), the House of Lords held that an occupational disability pension whether or not contributory is not deductible. And in *McCamley* v. *Cammell Laird Shipbuilders Ltd*

(1990) the proceeds of a personal accident policy taken out by the employers for all their employees and financed by the employers were held non-deductible. However, contractual sick pay must be deducted from damages for loss of earnings.

7.16 Fatal accidents

Relatives of the deceased worker can claim for the loss of their breadwinner caused by the defendant's negligence or breach of statutory duty if dependent on him at the date of death (Fatal Accidents Act 1976). Spouses, former spouses, children, parents, grandparents, grandchildren and siblings, step-parents and children, uncles and aunts and their issue, and adopted and illegitimate children are all included, as is any person who has lived in the same household as husband or wife of the deceased for at least two years before the death. The action is for financial loss caused by the death, not for grief and sorrow, except that spouses and parents of unmarried children under 18 are entitled to £7,500 for bereavement. The introduction of this head of damages by the Administration of Justice Act 1982 has caused resentment and misunderstanding in cases where young children have died, because the parents have had the impression that their child is being valued as worth only the statutory amount. It is a truism that it is cheaper to kill than to maim: no damages need be paid to compensate a dead plaintiff for the years of life or earnings lost, other than to the extent that others would have been financially dependent on him.

In assessing damages, no account is taken of any property inherited from the deceased, nor of any insurance monies. Social security benefits received by the dependants are not deducted. No reduction will be made in a widow's damages for wages she can earn; if she was not working during her husband's life she can claim full loss of dependency.

7.17 Exemplary damages

In the United States, damages for personal injuries are much higher than in the UK, partly because juries and not judges fix the amounts, but also through the availability of exemplary (punitive) damages. In English and Scottish law, the purpose of damages in most cases is to compensate, not to punish. The latter is the function of the criminal law. In many American jurisdictions, the employer who negligently exposes his workforce to some dangerous substance will have to pay not just for their loss of earnings, medical expenses etc. Damages will be increased by a substantial amount to demonstrate the jury's disapproval of the employer's conduct. In our philosophy, it is wrong to use damages as a means of punishment when it is not the wrongdoer but his insurance company who pays, but there may be an element of hypocrisy here since we do give damages for non-pecuniary loss like pain and suffering, which many regard as a concealed penalty.

In *AB* v. *South West Water Services* (1993) the plaintiffs suffered ill-effects as the result of drinking contaminated water polluted by the introduction of 20 tonnes of aluminium sulphate into the system at a water treatment works. It was held that only compensatory, not exemplary, damages would be awarded.

7.18   *Limitation of actions*

In the civil law, a person who wishes to claim a remedy from the courts must start his action within a period of time fixed by statute. There is a number of reasons for this. The memories of witnesses will fade with time, the defendant and his insurance company must be allowed after a reasonable period to proceed with their business on the assumption that the plaintiff is not going to pursue the case, and it is in the general public interest to limit litigation. Different periods are specified for different purposes. In unfair dismissal, the dismissed employee only has three months from dismissal to start proceedings in an industrial tribunal, whereas a plaintiff suing for property damage caused by a breach of a contract made by deed has twelve years from the date of the breach. Where such an action is based on the breach of a simple contract not under seal, the plaintiff has six years from the breach. Time stops running against the plaintiff on the day his writ is issued. There is then usually a very substantial delay before the case comes to trial, if indeed it ever does, for most actions are settled out of court. The court has power to prevent a plaintiff from continuing with his action if he is unreasonably dilatory in proceeding with it.

In actions for breach of contract or tort in respect of personal injury (including disease or impairment of physical or mental condition), the limitation period is normally three years from the date on which the cause of action accrued, that is, in contract the date of the breach and in tort when the damage was suffered

(Limitation Act 1980). In most cases, this is a simple matter to determine, because the plaintiff knows only too well the date on which he was injured. However, if the plaintiff has contracted a disease which only manifests itself years after his exposure to some dangerous substance, his three years may pass long before the symptoms appear. In one case, the plaintiff had contracted pneumoconiosis after years of exposure to noxious dust, but had not been exposed during the period immediately before the illness was diagnosed. If the limitation period was strictly applied, it was too late to bring an action for compensation against the employer who had negligently exposed him to the dust. To do justice in this type of case, Parliament provided that in a personal injury case the plaintiff has an alternative of bringing the action within three years of the date of his 'knowledge'. The knowledge in question is actual or constructive knowledge of all the following facts:

(1) that the injury in question was significant;
(2) that the injury was caused in whole or in part by the act or omission alleged to be wrongful;
(3) the identity of the defendant (or of a person for whom the defendant is alleged to be liable).

Thus, an employee who did not know the name of the company employing him because it was one of a group of companies all with similar names and he had been given the wrong one was able to bring his action within three years of discovering his employer's name (*Simpson* v. *Norwest Holst* (1980)).

If the plaintiff does not know these facts because he turns a blind eye, as where he refuses to see a doctor about obvious symptoms, he will be deemed to have that knowledge which he would have acquired had he sought expert advice. Suppose that he does seek medical help and the doctor negligently fails to diagnose his true condition? The Act states that he is not deemed to know what the expert should have told him but did not, so he will not be legally prejudiced by his doctor's negligence in such circumstances.

Since it is knowledge of facts and not law which is important, a plaintiff who knows that he is ill, knows the cause, and knows the name of the person who probably caused the damage, will lose his right of action after three years if he does not issue a writ because he is ignorant of the law. This is so even in a case where he has received bad advice from a lawyer. He can sue the lawyer for negligent advice but a lawyer, like a doctor, is not negligent just because he is wrong. However, the court has an overriding discretion to allow an action to proceed notwithstanding the expiry of the limitation period if it considers it equitable to do so. It must take into account the length of and the reason for the delay, the perishability of the evidence, the conduct of the defendant after the cause of action arose, including his response to the plaintiff's request for information, the duration of any disability of the plaintiff arising after the cause of action, the extent to which the plaintiff acted promptly and reasonably once he knew the facts, and the steps taken by the plaintiff to obtain medical, legal or other expert advice and the nature of any advice received. Judges tend to by sympathetic to those who refrain from suing at first because they want to continue working.

In *Brooks* v. *J. and P. Coates* (1984), the plaintiff worked in the defendants' cotton spinning mills in Bolton from 1935 until 1965. In the course of his employment he was exposed to large quantities of fine cotton dust. During that period fine cotton dust was not regarded by the experts as dangerous to health. Brooks, who smoked 10–20 cigarettes a day, left his employer in 1965 because of breathlessness and bronchitis which he and his doctor thought was exacerbated by the dusty conditions at work. In 1979, by chance, he had a talk with a friend who had worked in a quarry and had been granted a disablement pension for silicosis. The plaintiff applied to the DHSS for a similar pension and was diagnosed for the first time as suffering from byssinosis. His disability was assessed as 40 per cent. In 1980, the plaintiff issued a writ against his former employers claiming breach of statutory duty and negligence.

Obviously, the plaintiff's action was *prima facie* out of time. He had had full knowledge of all the relevant facts since at least 1965. He may not have known the name of his illness but he knew that he had symptoms and that these might be caused by cotton dust. The plaintiff therefore asked the court to exercise its discretion in his favour. The judge decided that Brooks was not blameworthy in not realising sooner that he had a legal claim and that, despite the disadvantage to the defendants of having to find evidence (both the mills where Brooks had worked had closed ten years before, records had gone and the workforce had dispersed), the prejudice to the plaintiff of being denied the right to litigate outweighed the prejudice to the defendants. The judge went on to decide that on the evidence Brook's chest problems were caused at least in part by cotton dust. This was somewhat surprising, considering the evidence that fine cotton dust rarely causes byssinosis and because the diagnosis of byssinosis depends on the plaintiff's history, particularly his own account of increased breathlessness on Mondays after the weekend absence, so that there is always a possibility of invention. It was held that the employers were not negligent because they could not have foreseen byssinosis in a fine cotton mill. They were, however, in breach of statutory duty under the Factories Act 1961 which obliged them to take 'all practicable measures' to render harmless all fumes, dust and other impurities as might be injurious to health (s. 4) and to protect against the inhalation of 'any substantial quantity of dust of any kind' (s. 63). It was practicable (defined as 'a precaution which could be taken or undertaken without practical difficulty') to control the dust by use of a vacuum process. The employers were held liable to the plaintiff, but his damages were reduced by 50 per cent for his cigarette smoking.

Where an employer has gone into liquidation before the employee brings his action the Companies Act 1989 permits the revival of the employing company in order that the employee may be able to claim on the employer's liability insurance policy.

These provisions in respect of latent damage were for a long time only applicable to actions for personal injury. After a series of test cases involving defective buildings which were discovered to have been negligently constructed years after completion, similar rules have now been introduced for negligence causing loss of or damage to property by the Latent Damage Act 1986. The plaintiff has six years from the date the damage occurred or three years from the date he has knowledge of all the relevant facts. There is, however, a long stop period of fifteen years from the date of the last act of negligence to which the damage in respect of which compensation is claimed is alleged to be attributable.

If the plaintiff is under 18 or of unsound mind at the time the right of action accrues, time does not start to run against him until he becomes an adult or recovers his sanity. This does not mean that no action may be brought meanwhile: procedures exist to allow minors and insane persons to bring actions through representatives.

7.19 Alternatives to the tort-based system

The increase in the numbers of civil actions for medical negligence has led to the medical profession to call for the introduction of a no-fault system of compensation, similar to those in Sweden and New Zealand. They are not alone. Accountants, architects, designers and solicitors are appalled at their potential liability and the costs of professional liability insurance. There is confusion about the exact definition of a no-fault system. It contains two central concepts:

(1) liability for damage without the necessity of proving negligence (often termed strict or absolute liability);
(2) a guarantee to the person injured that he will receive the compensation to which he is entitled because the money comes, not directly from the wrongdoer, but from a central fund.

In the field of industrial injuries, of course, we already have a no-fault system. It is based on a relatively simple administrative process. In 1980 the DSS estimated that the costs of running the scheme were only 13.3 per cent of the combined total of compensation and operating costs. The Pearson Commission in 1978 estimated that each pound in tort damages took at least 85p in costs to obtain. Atiyah has written in his major work on *Accidents, Compensation and the Law*:

'It is hard to believe that anyone could make a dispassionate review of the tort system and the industrial injury system, without coming to the firm conclusion that on almost every count the latter is a superior and more up to date model of a compensation system.'

Proposals to extend the industrial injury scheme to, for example, commuting accidents or individual cases of disease proved to have an occupational cause, are increasingly countered by the objection that it is fundamentally unjust to discriminate by paying more to the industrially disabled than to those disabled by non-industrial illness or injury. However, the introduction of the Criminal Injuries Compensation scheme, discussed in Chapter 5, compensation for damage caused by mass vaccination of children (Vaccine Damage Payments Act 1979), compensation for haemophiliacs who contracted AIDS from contaminated blood products, all three financed by the State, and strict liability on the producer of defective products (Consumer Protection Act 1987), shows that governments are conscious of the disadvantages of a negligence-based system, though the Pearson Commission's recommendation of a trial scheme for road accidents, financed by a special levy on the price of petrol, has never been implemented. Would a system of liability without fault not be preferable in all cases of personal injury?

One misconception easily dispelled is that no-fault liability does away with the need for lawyers. Strict liability for vaccine damage still obliges the plaintiff to prove that brain damage was *caused* by the whooping cough vaccine. Under the Consumer Protection Act, it is still for the plaintiff to prove that the product was *defective* and to find some defendant who is responsible. If costs are to be kept in check, any system of no-fault compensation for medical injury would probably have to distinguish between the consequences of 'natural' ageing and disease, and injuries which should not have occurred in the normal course of events.

Who would benefit from the general introduction of a no-fault system? All those potential claimants who now fail to recover damages in tort because they are unable to prove negligence and those who now cannot obtain compensation because the person responsible for their injury has neither money nor insurance.

Should the tort system run parallel with the no-fault system, as at present in the field of industrial injury, or should it be superseded? An advantage of retention might be that the tort system could be used to top up a basic State level of compensation. On the other hand, the costs and inadequacy of the tort system cry out for reform. The Pearson Commission thought that tort was a method of holding wrongdoers personally responsible and was therefore a deterrent, but it can be argued that the criminal law and professional disciplinary bodies like the General Medical Council and the UK Central Council are sufficient and potentially more effective in this respect.

Finally, how should any new system be financed? The industrial injuries scheme is administered by the Government and paid for by contributions levied on all employers and employees equally. Product liability is in the hands of individual producers and their insurance companies. In the present political climate, it is more likely that the Government would favour a system based on private insurance. An important issue is whether there should be varying premium rates based on the degree of risk, unlike the industrial injuries scheme, but similar to employers' liability insurance. The theory of general deterrence developed in the United States by Calabresi and others holds that:

> 'If we can determine the costs of accidents and allocate them to the activities which cause them, the prices of activities will reflect their accident costs, and people in deciding whether or not to engage in particular activities will be influenced by the accident costs each activity involves.'

In other words, an employer will be deterred from using a dangerous process because the costs of insurance put up his operating costs and result in less profit. The evidence from many countries is that the introduction of workmen's compensation laws dramatically reduces the numbers of industrial accidents. It is submitted, however, that market forces are not the only relevant considerations. Obstetricians and orthopaedic surgeons are more likely to be sued for negligence than any other doctors, but this is not an argument for more Caesarian sections and fewer hip replacements. Even in commercial industry, the idea that if the profit is high enough a risk becomes acceptable has a decidedly nineteenth century ring about it.

Chapter 8

Employment Law

8.1 Introduction

The law relating to employment is concerned with the relationship between the employer, his employee and the employee's trade union. As with the law of industrial injuries, the sanctions provided for breach of the rules are in the form of money compensation, but there is also the possibility of court orders protecting property rights or ordering a return to legality. An injunction is a powerful remedy, usually obtained from a High Court judge, disobedience to which is a contempt of court for which the contemnor can be fined or sent to prison.

Employment rights are regulated partly by the law of contract and partly by statute, of which the most important is the Employment Protection (Consolidation) Act 1978 (EPCA). Other legislation is comprised in the anti-discrimination laws enacted in the Equal Pay Act 1970 (EPA), the Sex Discrimination Acts 1975 and 1986 (SDA) and the Race Relations Act 1976. European law has been of particular significance in the employment field, as explained in the introduction. Significant changes, several stemming from Europe, are incorporated in the Trade Union Reform and Employment Rights Act 1993 (TURERA). Actions for breach of contract must be pursued in the High Court or County Court, where it is usually necessary to be represented by a lawyer and legal aid is available, whereas the statutes confer jurisdiction on the industrial tribunals, informal courts chaired by a lawyer who sits with two lay persons, one representing employers and the other employees. Legal aid cannot be obtained to take a case to an industrial tribunal, but it is not necessary to be represented by a lawyer, and trade union officials and personnel managers regularly appear. The loser before the tribunal does not have to pay the winner's costs (unlike in the High Court and County Court) unless he has brought or defended the case unreasonably. Appeal lies from the tribunals only on a point of law to the Employment

Appeal Tribunal, which sits in London and Glasgow and which is composed of a High Court judge and two lay representatives from both sides of industry. Legal aid is available in this court. Further appeals may be taken to the Court of Appeal and the House of Lords.

If an employee is dissatisfied with his employer's behaviour but wants to stay in his job, he may sue for breach of contract of employment in the County Court. The alternative, when the employee can tolerate no more, is to resign and claim constructive dismissal before an industrial tribunal. This has the major disadvantage that there is an upper limit to compensation in the industrial tribunal (in 1993 £11,000), the average level of damages awarded is far lower than this, many tribunals reduce damages by a percentage because of the employee's contributory fault, and the success rate of applicants to industrial tribunals is only about 30 per cent. Tribunals have power to recommend reinstatement, but cannot force an employer to re-employ a worker who has been dismissed.

A very important institution in the employment field is ACAS, the Advisory, Conciliation and Arbitration Service. Although paid for by taxation, this body is independent of Government, having its own Council with representatives from both sides of industry and independent members. When an application is made to an industrial tribunal ACAS will offer its services to both parties to try to conciliate. The majority of applications are dropped before they reach the tribunal partly because of the effectiveness of this procedure. ACAS also gives advice about industrial relations both to companies and individuals and arranges arbitrations in industrial disputes by agreement of both parties. The award of the arbitrator is not usually legally binding, though it is open to both parties to make it binding by agreement in writing. A 'pendulum' arbitration is one in which the parties have directed the arbitrator to decide either for one side or the other, but not to split the difference between them.

The bulk of the work of industrial tribunals is concerned with allegations of unfair dismissal. The introduction of the unfair dismissal law by the Industrial Relations Act 1971 has had significant effects on the conduct of industry. Now that an employer who dismisses an employee must give a reason and be prepared to defend it at a tribunal, employers have refined their disciplinary procedures and created new ones. Justice demands that the employee be allowed to speak in his own defence before action is taken against him, and that he should not be dismissed for misconduct (other than gross misconduct), incapability or even ill-health, unless he has previously been warned that this is a likely outcome if there is no improvement. Employers have rewritten their disciplinary rules and agreed new procedures for dealing with conflict, providing for a system of informal and formal warnings, and appeals to a higher level of management. Employers feel that unmeritorious claimants use the tribunals for their nuisance value. There has been a steep decline in the use of pre-hearing assessments whereby the tribunal chairman may warn either party that he has no case, and that if he proceeds and loses he will be ordered to pay at least part of the costs of the other side.

Occasionally, the High Court procedure is used to try to obtain an injunction. The employee who has been dismissed in breach of the agreed procedure may ask to be reinstated while his appeal is heard. An employer may ask the court for an injunction prohibiting the employee from publishing confidential information belonging to the employer. Such remedies can never be obtained from an industrial tribunal.

The occupational health professional may be asked to give evidence before one of these bodies. His duty of confidentiality directs him to refuse to give clinical information about the employee without the employee's consent to ACAS or at an internal interview or appeal, unless there is some overriding public interest, though he can be compelled to answer at a court or tribunal. Sometimes, his care for his patient will conflict with the employer's wish to dispense with an unsatisfactory employee. In such circumstances, he must seek as an expert to give information to both sides without himself trying to make the decision about the employee's future.

8.2   The contract of employment

A contract is made when one party makes an offer to another and he accepts. This need only exceptionally be in writing (e.g. contracts for the sale of land in England). Contracts of employment must be distinguished from other legal relationships. As has been explained in Chapter 2, the employment relationship is marked by the employee's subjection to his employer's control and participation in his business organisation, rather than being in business on his own account.

It is often hard to identify the terms of a contract of employment. Express terms may be found in letters of appointment, written contracts or remarks made at an interview, for example, but many contract terms are never put into words but are included by implication. Often, parties do not express their intentions because they assume that they are obvious. In one case a steel erector working for a civil engineering company refused to move to a site too far from his home to allow him to live there in the week. The court held that the nature of the business was such that employees impliedly agreed to be mobile (*Stevenson* v. *Teesside Engineering Ltd* (1971)). In 1963 an Act of Parliament was passed with the intention of protecting the employee by imposing a legal obligation on his employer to put the terms of the contract into writing and give him a copy so that he knew where he stood. Because that statute was called the *Contracts of Employment Act*, these statements are universally known as contracts of employment but this is a misnomer. The written statement is evidence of what the employer says was agreed, but it can be challenged as inaccurate. In practice, it usually goes uncontradicted and is important evidence when a dispute arises, sometimes years later. The signature of the employee makes no difference to its binding force unless the employee has signed an express agreement that the statement incorporates his terms and conditions of employment. The statutory obligation, now to be found in the EPCA, as amended by the Trade Union Reform and Employment Rights Act 1993, is to give a written statement within eight weeks of the employee starting work. It does not extend to all the terms of the agreement, but to pay, hours, holidays, absence due to sickness or injury, sick pay, pensions, notice, job title, place of work and details of disciplinary and grievance procedures (but not those relating to health and safety at work). The statement should inform the employee of any collective agreements which directly affect his contract. The employer must update the statement within one month of any change in the contract taking effect. If the employer fails to deliver the statement the employee can complain to an industrial tribunal which can remedy the defect.

Many employees have their contractual terms regulated by negotiations between trade unions and employers in the process called collective bargaining. The collective agreement between union and employer is not legally binding on those parties unless it expressly states in writing that it is intended to be so (Trade Union and Labour Relations (Consolidation) Act 1992). There are a few single union agreements which incorporate such a clause, but they are very much in the minority. But even a non-binding collective agreement becomes legally enforceable if it is incorporated into the individual contract between employer and employee. Thus, if an employee is notified that in the event of sickness absence the collective agreement for the industry entitles him to up to six months full pay, he can sue the employer for breach of contract when he does not comply.

English law is remarkable for the degree to which it still leaves regulation of the employment relationship to the agreement of the parties. There is no minimum wage fixed by law since the Wages Councils were abolished in 1993. Nor is there any minimum holiday provision. Such a system has allowed the judges scope for judicial engineering in the erection of a structure of implied terms. The duty of trust and confidence has been discussed in Chapters 2 and 3. In *Scally* v. *Southern Health and Social Services Board* (1991) employees of the Northern Ireland health boards were required to make contributions to a statutory superannuation scheme. Full benefits depended on the completion of 40 years' service. Regulations were introduced in 1974 allowing employees to purchase extra years of service in order to make up the 40 years' contribution, but this right had to be exercised within 12 months of the coming into force of the regulations. Four doctors failed to take up the option in the regulations because their employer had never brought it to their attention. The House of Lords held that there is an implied obligation on the employer to take reasonable steps to publicise this kind of contractual benefit in order that the employee may take advantage of it. Substantial damages were awarded.

There is, however, no implied term that the employer will provide personal accident insurance for an employee who works abroad, or advise the employee on the need to take out his own insurance (*Reid* v. *Rush and Tompkins Group* (1989)).

There is an implied term in every contract of employment that the employer will take reasonable care to protect his employee's safety and health while he is acting in the course of his employment. Many examples could be given. One is that of Mrs Austin who worked for BAC in a job where she had to wear protective eyewear. She asked a manager on a number of occasions whether it would be possible to provide her with protective glasses made up to her own prescription because she found it cumbersome to wear the protective spectacles over her own glasses. The manager brusquely rejected her complaint and refused to investigate the matter. It was held to be a breach of contract by the employer not at least to investigate reasonable safety requests (but not frivolous complaints) (*BAC* v. *Austin* (1978)). Mr Justice Phillips said this in the course of his judgment:

'It seems to us that it is plainly the case that employers ... are under an obligation under the terms of the contract of employment to act reasonably in dealing with matters of safety, or complaints of lack of safety, which are drawn to their attention by employees ...'

Another case concerned Mrs Firth (*Oxley Steel Tools* v. *Firth* (1980)). She had to endure working for several months in intolerably cold conditions: a breach of contract by the employer as well as a breach of the employer's statutory duty under the Factories Act.

The importance of implied terms leads to the question of how they may be identified. There are basically three kinds. The first is established by custom and practice in the industry. This must be 'reasonable, certain and notorious', so well known that everyone in the field takes it as read. The second is created by the previous behaviour of the parties. If Joe Bloggs has worked for the same employer for three years and has throughout that time accepted that he will work an extra hour each Friday it is likely that this will have become contractually binding. The third is created by the judges from the employment relationship. The courts have held that schoolteachers are under an implied duty to fill in for colleagues off sick (*Sim* v. *Rotherham MBC* (1986)).

Express terms cannot in general be contradicted by implied terms. If you want the position to be clear you should make an express agreement, preferably in writing. Even then, the court may use its power of interpretation to vary the effect of the term. In *United Bank* v. *Akhtar* (1989) there was an express clause allowing the employer to move the employee to any branch of the bank in the UK. The employer ordered Akhtar to move from Leeds to Birmingham at only six days' notice, refusing to consider his personal circumstances. The Employment Appeal Tribunal decided that the employer was in breach of an implied term in the contract to show trust and respect to his employee.

Neither party to a contract of employment may unilaterally change its terms without the other's agreement. However severe the employer's financial difficulty, he cannot reduce wages without consent. Confusion sometimes arises between the specific contractual duties of the employee and his general implied obligation to obey his employer's lawful instructions. Contract terms cannot be changed by the employer alone, but managerial prerogative can decide on working rules and practices.

Can management lawfully introduce a no smoking rule and impose it on employees without their consent? The answer is that the employer's duty to protect his employees and the employee's duty to co-operate entitle the employer to introduce new health and safety rules without agreement, just as he can unilaterally change disciplinary rules. Ms Dryden was employed as a nursing auxiliary in the theatre section of the Western Infirmary, Glasgow. She was accustomed to smoking thirty cigarettes a day. Her job was such that she was unable to leave the premises during the course of the day but, until 1991, areas were set aside within the employer's premises in which smoking was permitted. The employers decided to prohibit smoking in all general and maternity hospitals. During 1991, all employees were issued with letters giving notice of the change with offers of advice and counselling. The policy was implemented in July 1991 and twelve days later the employee resigned, claiming constructive dismissal. The Employment Appeal Tribunal held that the employers were not in breach of contract. There was no implied term that the employee would be permitted to smoke at work. Ms Dryden lost her case (*Dryden* v. *Greater Glasgow Health Board* (1992)).

To comply with the law of unfair dismissal and, more important, to be likely to achieve general acceptance, the imposition of a ban on smoking should usually be enforced only after consultation with the workers and consideration of what is reasonable. The employer must be able to prove that he has a good reason for the change. He may be able to point to a fire risk or the dangers of passive smoking to non-smokers. Here, the health education role of the OH professional will come to the fore. It is possible that the medical evidence about passive smoking now *obliges* the employer to ban smoking in order to comply with the statutory duties under the Health and Safety at Work Act (Chapter 5). The provision of a smokers' room, if feasible, should be considered. The ban must be clearly communicated to the workforce individually (not merely through the trade union), together with the consequences of breach. In most cases advance notice should be given, because a reasonable manager recognises that a smoking habit is not easy to relinquish. In a tribunal case, the employers introduced a ban on smoking after a fire, notified each employee individually and posted notices in the no smoking areas. An employee was dismissed when found smoking in the toilets, where smoking was forbidden. The tribunal decided that the dismissal was fair (*Martin* v. *Selective Print* (1987)).

In contrast, because of the principle of inviolability of the person, I think it unlikely that English courts would imply an obligation to submit to body searches, random testing for drugs or alcohol, other biological testing or medical examination without the employee's agreement, unless the special nature of the job allowed the implication of such a term. Health professionals may have a term implied in their contracts that they will undergo medical examination if patients may be at risk (*Bliss* v. *SE Thames RHA* (1985)) (see Chapter 4).

In any event, developments in the law of unfair dismissal discussed later in this chapter mean that an employer may be held to have fairly dismissed an employee who refuses to agree to a change in the terms of his contract if the employer can show a good reason (like the protection of other employees from passive smoking) for imposing the change.

8.3   Contractual sick pay

The importance of making clear the terms of a contract can be illustrated from the legal rules about the employee's right to payment when he is absent through ill-health. Originally, the rule was 'no work – no pay'. Then, the Welfare State provided social security benefits for those who could not work because of ill-health, financed by contributions from employers and workers. Trade unions on behalf of their members negotiated collective agreements with employers whereby they agreed to 'top up' sickness benefits to provide normal earning levels for at least a period (some 80 per cent of workers were covered by such schemes by 1975). These were incorporated into contracts of employment and became express contractual terms. What of those who did not have the benefit of an occupational sick pay scheme? One such was Mr Mears, a security guard, working for a small family business. There was no provision in his written statement about sick pay and nothing had been said. There was evidence that it was not the practice of the company to pay employees when they were off sick and that this was well known to them. It was held that the employer was not bound by an implied term to pay contractual sick pay (*Mears* v. *Safecar Security* (1982)).

8.4   Statutory sick pay (SSP)

Soon after the Mears case was decided, an Act of Parliament transferred the obligation to provide income support during short periods of sickness from the DSS to the employer under the

Statutory Sick Pay scheme, now to be found in the Social Security Contributions and Benefits Act 1992. Now, the employee is entitled to payment of at least part of his wages by his employer for up to 28 weeks. The costs of administering the scheme fall mainly on the employer who is responsible for all short-term sickness benefit and recovers 80 per cent of the money paid out by deducting it from his National Insurance contributions (small employers may claim 100 per cent rebate). The employee is liable to pay income tax on sick pay. SSP will not give the employee the whole of his wages during sickness: he will have to prove a contractual obligation on his employer to make the amount up to full pay. Note that the same rules apply whether a worker is unable to work because of a work-related accident or industrial disease or because of an illness totally unconnected with his work. No statutory payment is due for the first three days, but many employers are contractually bound to make payment for that period. Part-time employees are included, and there is no qualifying period of employment, but employees over State pension age, those employed for a fixed term of three months or less, and those with average earnings of less than the lower earnings limit for National Insurance liability are excluded.

8.5   Proof of sickness

Before 1982, an employee who was off work for more than three days needed a medical certificate from his GP to claim sickness benefit. The same certificate was used to claim contractual sick pay from the employer. This system was notoriously unscientific. In 1982 the (then) DHSS adopted a procedure of self-certification for the first seven days of illness and the GPs' contracts with the Family Health Services Authorities were changed so that they were no longer obliged to give a free medical certificate for an absence from work of less than seven days. In practice, therefore, most employers now rely on self-certification for the first seven days for all purposes. The SSP rules provide that an employer is not entitled to ask for a doctor's statement for the first seven days of a spell of sickness. If a doctor states that an employee should not work for precautionary reasons, or because he is convalescing, the employee can claim SSP despite not being ill (Statutory Sick Pay (Medical Evidence) Regulations 1985).

8.6   The control of absenteeism

This is a problem which affects contractual sick pay, SSP and may also be relevant in disciplinary proceedings. It is a management not a medical problem with which each employer must deal for himself in consultation with worker representatives. Employers are advised not to challenge a doctor's note unless they have clear evidence to the contrary. In *Hutchinson* v. *Enfield Rolling Mills* (1981), the employee, a maintenance electrician, was diagnosed by his GP as suffering from sciatica and signed off for seven days. Two days later, he was seen by one of the directors taking part in a union demonstration in Brighton. He was dismissed for gross misconduct after consultation with the company's doctor. The Employment Appeal Tribunal disagreed with the industrial tribunal's refusal to go behind a sick note:

'The employer is concerned to see that his employees are working, when fit to do so, and if they are doing things away from the business which suggest that they are fit to work, then that is a matter which concerns him.'

This does not mean that an employee while off sick must remain at home in bed at all times. It might well be that an employee convalescing from a serious illness, but still unfit to work, could without any misconduct be sunning himself on the front at Brighton.

Those suspected of abuse of self-certification may be refused SSP; they may then appeal to a DSS adjudication officer. Medical advice may be sought from the occupational health service or the GP, but the employee's consent will be necessary. It is advisable for employers to anticipate these matters in the contract of employment and include contractual obligations to submit medical reports after periods of absence. If an employee has self-certified himself four times within a year, the DSS suggests that the suspicious employer may refuse to pay SSP on the fifth occasion and send him to the DSS for medical checks. If the employer can prove that the employee lied on the certificate, this may be treated as misconduct and give rise to disciplinary action, but action of this kind based on mere suspicion without proper investigation can amount to a constructive dismissal. A self-certification case was *Bailey* v. *BP* (1980). Bailey, a rigger at an oil refinery, went on holiday to Majorca and, on return, certified that he had been suffering from a 'gastric stomach'. Unfortunately for him, he had been seen in Majorca by the assistant maintenance engineer. His summary dismissal was upheld as fair by the Court of Appeal.

It is important that employers keep full records of sickness absence. Patterns of ill-heath are often the best way to identify an employee with a drink problem ('the Monday morning syndrome') who may be persuaded to undertake treatment at an early stage. Many employers set absence norms for the workforce and provide that any employee who exceeds the norm will be investigated, possibly including an interview with the company doctor or nurse. Medical advice may also be indicated if an employee has been continuously absent for some weeks (Gardner (1983)). These procedures should be communicated to each employee and incorporated into individual contracts of employment. An employer with an OH service is at an advantage, because he will have expert knowledge not just of medical aspects but also of the nature of the work. As the ACAS Advisory Booklet, *Absence*, puts it:

'The company doctor becomes familiar with the kind of work done, and the typical stresses caused by it. He is better able to judge whether someone who is sick or 'off colour' can safely be allowed to continue to work, or should be sent home to avoid accident or more prolonged illness. Serious illness may be spotted at an early stage.'

8.7 Notification to the employer

Well-drafted company rules will impose on the employee the duty personally and as early as possible to let the employer know by telephone that he is unable to come to work. This may be made a contractual obligation, breach of which may lead to disciplinary action. The SSP rules allow an employer to withhold SSP as a penalty for late notification, but only within limits laid down in the regulations:

(1) Notification cannot be required by a specific time on the first qualifying day. Any time during that day is sufficient.
(2) The employer must accept notification by some other person.
(3) The employer cannot demand a medical certificate as notification and he cannot require notification more frequently than once a week during the sickness.

The employer does not have to change his company rules about notification of sickness, but if they are strict he may find that a notification late by company standards and thus constituting a breach of discipline is in time for SSP.

8.8 *The employer's right to suspend*

Sometimes, the employee is willing to work but is prevented by the employer. Can the employee demand compensation for being deprived of the opportunity to earn his wages? It depends on the nature of the contract and the reason for the suspension. As a general rule the employer has a contractual duty to pay wages while the contract of employment subsists. The secretary who comes to work and is given nothing to do must still be paid. As long as the employer pays wages he is usually not obliged to provide the employee with work, so suspension on full pay at least for a short period is lawful. There may, however, be either express or implied terms in the contract which allow the employer to suspend without pay. In some companies, suspension without pay is used as a method of discipline according to a disciplinary code. This is lawful if the code has been properly notified to the employees, usually when they are given their written statements at commencement of their employment. In some industries, like construction or ship-building, there is an implied term in the employees' contracts that the employer can lay the workers off temporarily because of bad weather or lack of orders but he is then often bound by collective agreements to pay a guaranteed minimum. There is also a limited statutory right to guarantee payments in the EPCA (up to twenty days a year), and unemployment benefit under the social security system, but unemployment benefit is not payable for any day for which the employee receives a guarantee payment.

8.9 *Medical suspension*

Health and safety legislation which provides for medical surveil-lance of those working with hazardous substances like lead or ionising radiations or those substances specifically mentioned in Schedule 5 of the COSHH Regulations gives EMAS or the Appointed Doctor the power in effect to direct the suspension of an employee on medical grounds. If the employee is suspended because he is incapable of work by reason of disease or bodily or mental disablement he will only have the right to SSP, unless in addition he has a contractual right to sick pay. If, however, he is suspended as a preventive measure so that he does not fall ill he will be entitled to remuneration under section 9 of the EPCA, amounting to full pay for up to 26 weeks. The employee must have been employed for at least one month at the time of the suspension and if he unreasonably refuses suitable alternative work offered by the employer, whether or not it is his usual job, he will lose his right to remuneration.

In a 1987 case, two men worked for a company manufacturing batteries, of which lead is an important component. They were subject to regular medical surveillance under the Control of Lead at Work Regulations 1980, which stipulate that if the blood lead concentration rises to 80 μg per 100 ml or above the Appointed Doctor or Employment Medical Adviser must in most cases certify the employee as unfit to work with lead, except where he has worked with lead for 20 years or at least ten years if he is over 40, in which case the AD or EMA has a discretion. Mr Appleton fell into a skip containing lead paste. His blood level was over the statutory level for a while, but even when it fell below that level, the GP did not consider that Appleton was fit for work. Another employee, Mr Hopkinson, was also kept off work by his GP because of high blood levels, but neither man had a certificate from an AD or EMA. Both men felt so unwell that they were unable to do any alternative work. The Employment Appeal Tribunal held that neither employee was entitled to medical suspension pay, though they could claim SSP. They were unable to do *any* job: medical suspension pay is intended for workers who are moved from a job carrying a particular risk to another, not those who are unable to work at all because of ill-health (*Stallite Batteries* v. *Appleton and Hopkinson* (1987)).

It is important to note that in a case where an employee complains that he has been unfairly dismissed for a reason which could lead to medical suspension, the minimum qualifying period of employment is one month, instead of the usual two years.

8.10 Changes in contractual terms

Once a contract is made, neither party can unilaterally alter its terms without the other's agreement. Management cannot introduce new working hours or payment systems or AIDS tests without the consent of each individual employee. Trade union officials have no power to give consent by proxy. The effects of this rule have been blunted by two legal developments. The first is the use by the judiciary of the implied term. An employee refuses to move from one site to another and is held to have impliedly agreed to be mobile. Another objects to the introduction of a new disciplinary code and finds that he has impliedly agreed to obey his employer's reasonable commands. Miss Glitz was appointed as a 'copy typist/ general clerical duties clerk' in a small office. Two years later she was told that she would have to operate the duplicator which she had not done before. She found that the vapour from the machine gave her headaches, though it was in good working order. The Employment Appeal Tribunal held that the operation of the duplicator was impliedly part of her job and that it was fair to dismiss her because of her incapability, since there was no other suitable job available (*Glitz* v. *Watford Electric Co. Ltd* (1979)).

The second is the fact that an employer almost invariably has the contractual right to terminate a contract of employment by giving an agreed period of notice. The courts have held that an employer who dismisses an employee with notice because he refuses to accept a change in his contract is not liable for unfair dismissal if he has a good business reason for making the change. A health authority wished to close down a small unit and move the employee to a different post in a larger one. This was a change in the employee's contract and he refused to move. The tribunal held that the employer had acted reasonably and was not liable for unfair dismissal (*Genower* v. *Ealing and Hounslow AHA* (1980)). Good industrial relations practice, of course, dictates that an employer must make every attempt at persuasion before reaching this stage. If the employee agrees to a change in his terms, the contract is varied by consent.

8.11 The right to dismiss

Employers' freedom of action is limited by the law of contract and the law of unfair dismissal. Dismissal in breach of contract is known by lawyers as wrongful dismissal. Actions for wrongful dismissal must be brought in the County Court if the damages claimed are less than £50,000 or in the High Court if higher amounts are sought. A dismissal is wrongful if the employer has broken a term of the agreement. Thus, if a company has agreed to employ a consultant for a fixed term of three years and dismisses him after six months it will have to compensate him for the loss of two and a half years' employment, minus what he can reasonably be expected to earn elsewhere. However, if the employee commits a serious breach of his side of the bargain the employer can treat the relationship as terminated without further obligation on his part: this is known as summary dismissal. Dishonesty usually justifies summary dismissal, whereas mere carelessness does not, even if the consequences are expensive.

The EPCA imposes minimum periods of notice on all employers who are free to agree longer but not shorter periods. Employees who have been employed for up to two years are entitled to be given at least one week's notice, and each extra year of employment gives the right to one more week's notice, up to twelve weeks. The employee must give a minimum of one week's notice, however long he has been employed. Wages in lieu of notice are permitted. Employees on fixed term contracts in which the employer has guaranteed employment for a set period cannot be dismissed by being given notice. Everything depends on the terms of the contract and it is advisable to make an express agreement in writing before the employment starts.

The employee who is dismissed with the agreed period of notice cannot usually complain that he has been wrongfully dismissed, unless he can show that a disciplinary procedure has been incorporated into his contract and that the employer has not observed it. The motive for the dismissal is irrelevant in the law of contract. For this reason Parliament in 1971 created the additional remedy of unfair dismissal which is concerned not with procedure, as is the law of contract, but with the reason for the dismissal. For the first time, the employer had to give the employee a reason and justify it as fair before an industrial tribunal if required by the employee to do so. Only employees under retirement age with a minimum continuous service of two years at the date of dismissal can complain to the tribunal. In March 1994 the House of Lords held that restrictions on part-time workers were contrary to European law, as indirect discrimination on the grounds of sex (most part-timers are female). An application to the industrial tribunal must normally be made within three months of the date of dismissal, that is, where notice has been given, three months from the date of the expiry of the notice, and where dismissal was without notice, three months from the date the employee left. The fact that an employee has been given proper notice does not prevent him from complaining that he was unfairly treated. Conversely, it is possible for a tribunal to hold that an employee sacked without proper notice was treated fairly in all the circumstances of the case (though this is less likely). Where a dismissal is for trade union membership or activities, or because the employee refuses to belong to a trade union, or is an act of sex or racial discrimination, no minimum length of qualifying service is needed by an employee who wishes to complain to a tribunal. A dismissal for a reason which could lead to medical suspension under s. 19 EPCA requires only one month's employment at the date of dismissal for the tribunal to have jurisdiction.

The introduction of unfair dismissal reduced the importance of wrongful dismissal, but actions for breach of contract are still brought in cases of high-earning employees because there is an upper limit on damages for unfair dismissal but not for breach of contract. Also, in exceptional cases, the High Court has power to order reinstatement, whereas the industrial tribunals cannot force the employer to take back the dismissed employee.

8.12   The concept of dismissal

An employee may be dismissed in any one of three ways. He may be told to go, with or without notice, he may be informed that the fixed term contract on which he is employed will not be renewed when it expires, or he may be the subject of 'constructive' dismissal. The employer dismisses constructively when he behaves in a way which constitutes such a breach of his side of the contract of employment that the employee is justified in terminating the relationship. An employer constructively dismisses an employee when he puts the employee's health or safety at risk. A manager of a shoe shop who failed to introduce any safety precautions, even to instal a telephone, after the shop had been robbed twice was held to have constructively dismissed one of the assistants who left after the second robbery (*Keys* v. *Shoefayre Ltd* (1978)). It is, of course, often a matter of debate as to whether the employer's conduct is bad enough to constitute a breach of contract on his part. The cases show that employers who fail to investigate complaints and show a complete lack of concern for the workers are likely to be held to have constructively dismissed. Mr Lynn, a manager who was the victim of such intemperate criticism by one of the directors of Weatherall's that he eventually resigned and succumbed to a nervous breakdown, was held to have been the victim of a constructive dismissal(*Weatherall Ltd* v. *Lynn* (1978)). At this stage it is important to emphasise a point which many misunderstand: a constructive dismissal, like dismissal with notice or summary dismissal, is not necessarily an unfair dismissal. A supervisor of a packing department who had been off sick was told that he would have to move to the job of production foreman. He refused because, he said, it would damage his health, but he would not agree to a medical examination to assess his suitability and he resigned. He had been constructively dismissed because the employer had no contractual right to move him to another job without his consent, but the employer had acted reasonably and was not liable for unfair dismissal (*Savoia* v. *Chiltern Herb Farms* (1982)).

If the employee voluntarily resigns from his job other than in response to his employer's unlawful behaviour, he is not dismissed and cannot claim any of the statutory rights dependent on dismissal, like compensation for unfair dismissal and statutory redundancy compensation. 'Voluntary' means what it says: an employee who is told that he will not be sacked if he agrees to go quietly is dismissed. Those who volunteer to be made redundant are usually held to be dismissed in law, but in a case concerned with the Universities' Premature Retirement Scheme the Court of Appeal decided that employees who accepted the generous compensation had terminated their contracts by agreement and had not been sacked. They were thus excluded from claiming redundancy payments (*Birch* v. *University of Liverpool* (1985)). Sometimes, it is in the worker's interest to wait to be told to go. A pregnant woman who gives up her job because she feels that the job is too much for her will run the risk of losing her statutory maternity rights, but if she is dismissed by the employer because she is unwell all her rights are preserved.

From time to time personnel managers make use of a legal doctrine quaintly known as 'frustration of contract'. The law provides that a contract automatically ceases, so that neither side has any further obligation under it, if it is overtaken by a supervening event that destroys its whole purpose. Frustration may be applied to the contract of employment in two separate situations: the employee's supervening illness or his incarceration for a criminal offence. The performance of the contract must have become impossible (as where the employee dies) or something radically different from what the parties contemplated when they entered into it. The principles to be applied by tribunals in an ill-health case were laid down in 1976 (*Egg Stores* v. *Leibovici*). Where illness or accident is relied upon as bringing about frustration, the employer must ask whether a time has arrived when he can say that matters have gone on so long and the prospects for future employment are so poor that it is no longer practical to regard the contract as still subsisting. Among the matters to be taken into account are:

(1) the length of the previous employment;
(2) how long it had been expected that the employment would continue;

(3)   the nature of the job (the employment of key workers is more easily frustrated);

(4)   the nature, length and effect of the illness and the medical prognosis;

(5)   the employer's need to appoint a replacement or to make a temporary replacement permanent;

(6)   whether wages or sick pay have continued to be paid.

The use of the doctrine in such cases is infrequent because of the effects on the employee. Imagine a worker who has just been seriously injured in a road accident or has had a crippling stroke. The employer writes to him telling him that, as he is unlikely to be able to work for at least a year, his contract is frustrated and he will receive nothing more from the employer, neither sick pay, nor wages in lieu of notice, nor compensation for the loss of his job. But the Court of Appeal in 1986 confirmed that the doctrine is still available to employers who wish to use it. Derek Notcutt had worked for a small company as a milling machine operator for 27 years. His contract of employment expressly provided that he had no right to sick pay (he would now be entitled to up to 28 weeks' statutory sick pay from the employer). When he was 63 he suffered a coronary infarction. When he had been off work for six months, the employers asked for a medical report from his GP. The latter wrote that he doubted whether Notcutt would be able to work again. The employers then gave him twelve weeks' notice, as they were advised was obligatory under the EPCA. The EPCA gives employees during the statutory periods of notice the right to full wages or to full sick pay even though if they were not under notice they would have no right to sick pay, so Notcutt claimed twelve weeks' sick pay. It was held that the heart attack had frustrated the contract of employment such that Notcutt's contract had automatically terminated. He was entitled to neither notice nor to pay (*Notcutt* v. *Universal Equipment* (1986)).

A contract of employment cannot be frustrated by the possibility that an employee's health may deteriorate. A manager employed as works director in a company in Darwen had a heart attack in April. He returned to work part-time in June and was told by a consultant that he would be able to resume full-time at the end of July. The employers felt that the job of works director would be too great a strain, so they offered him an alternative job at a lower salary which he refused. He was then given notice. The employers' decision was based not on any medical report, either from the employee's doctor or the company's own doctor, but on the statistical possibility that there might be a second heart attack. They said that the contract had been frustrated. It was held that frustration is brought about by events, not the risk of such events. There was a dismissal and it was unfair because it was not based on medical evidence about the employee (*Converfoam* v. *Bell* (1981)).

An alternative to dismissal of an employee who has had a long period of absence due to ill-health, and is unlikely to return in the near future, is to 'suspend' his contract, with his agreement. He is transferred to a 'holding department' until his return to work or his retirement, whichever is the earlier. The advantage of this scheme is that the employee, though not obliged to work and not in receipt of salary, maintains continuity of employment, because he has not technically been dismissed. Depending on the rules of the pension scheme, he may be able to accrue further pensionable service.

8.13   *The reason for the dismissal*

The employee has to prove that he has been dismissed. If he does so, the burden shifts to the employer to give a reason; the tribunal then decides whether in all the circumstances he acted justly and equitably in treating it as justifying dismissal. If the case goes to an industrial tribunal they only require to be convinced that the employer acted within a range of reasonable responses, not necessarily that the members of the tribunal would have reached the same decision. Only if the employer has made a decision so unreasonable that no reasonable employer could have made it can he be held liable for unfair dismissal.

The EPCA lists five possible groups of fair reasons:

(1)   capability or qualifications, capability being assessed by reference to skill, aptitude, health or any other physical or mental quality;
(2)   misconduct;
(3)   redundancy, that is that the employee's job will disappear;
(4)   that the employer cannot continue to employ the employee without contravention of a statute;
(5)   some other substantial reason.

In addition, the employer has to show that on the facts he acted justly. For example, an employer may give it as a reason for dismissal that the employee is dishonest, but if he has no evidence on which to base his suspicions he will be held to have acted unfairly. The Act provides that the size and administrative resources of the employer's undertaking are to be taken into account in deciding what was reasonable.

8.14 Dismissal for incompetence

The employer is not expected to continue to employ indefinitely someone whose performance in the job is poor. As there is a two-year qualifying period for unfair dismissal, employers are advised to review all their employees after eighteen months or so. Where the employee has worked for more than two years, the employer will have to be able to produce evidence of unsatisfactory work and to demonstrate that the employee has been made aware of his shortcomings and given time and, if necessary, further guidance or training to help him to improve. It is important to keep a record of interviews at which the employee was counselled and warned that improvement was necessary if the employee were to continue in employment.

8.15 Ill-health dismissals

This is a difficult area in practice because the employee may have to lose his job for something which is not his fault. Any member of a profession whose main purpose is to care for an individual will find it hard to be associated in any way with depriving a patient of his opportunity to work. On the opposite side of the coin are those who are suspected of malingering. If the employer asks the doctor or nurse to act as detective and prosecuting counsel, the confidence of the workforce in his impartiality is at risk. Only if the occupational health service is known to be both competent and fair will respect be maintained.

The leading case on dismissal for ill-health is *East Lindsey District Council* v. *Daubney* (1977). There, a surveyor employed by the Council was dismissed after long periods of absence due to 'anxiety and general debility'. The personnel director had sought medical advice from the district community physician, who wrote that the employee was unfit to carry out his duties and should be retired on the ground of permanent ill-health, and acted on it without indicating to the employee that his job might be at risk or allowing him to obtain his own doctor's report. The dismissal was held unfair. Tribunals must consider whether the employer acted reasonably in reaching a decision that he could no longer continue with the employee in an unfit state. Obviously, the two most important factors are the nature of the job and the nature of the employee's illness. The size of the employer's business is a relevant factor: it is easier for British Telecom to hold a job open for a year than it is for a small company. As a general rule it will be necessary to obtain medical evidence before making a decision, both to assess the potential for return to work and the likely capability of the employee if he is allowed to return. It is also advisable for a manager to interview the employee in person, or at least communicate with him by post, to ask how he feels about coming back and warn him that the employer is considering dismissal. It is not true that employers are prohibited from dismissing before the employee's entitlement to sick pay expires, nor should they automatically dismiss as soon as that period is over. If it can be shown that an employer's main reason for ending his employee's contract was to avoid paying statutory sick pay, he is legally required to go on paying until his liability ends for some other reason, and the employee might well be able to claim unfair dismissal in such circumstances.

One employer had a contract with employees which provided that a worker's absence for ill health or injury of 225 or more working days in the preceding 12 months entitled the company to give notice of dismissal. An employee, Brookes, had suffered from a recurrent depressive illness for 20 years. At the time he was eventually dismissed, he had not been absent for 225 days. The Employment Appeal Tribunal held, nonetheless, that the dismissal was capable of being fair, depending on the medical evidence and all the other facts of his case (*Smith's Industries* v. *Brookes* (1986)). Equally, a dismissal after an absence of more than 225 days could have been held unfair.

The purpose of consultation with the employee is partly to enable the true medical condition to be ascertained. It is also to allow the whole employment situation to be assessed and to consider whether the employee could be offered ill-health retirement or a different job more suited to his condition. The fair employer should at least consider the availability of other posts, but it has been held that he is not obliged artificially to create work for a sick employee, however deserving his case. In *Garrick's Ltd* v. *Nolan* (1980) a maintenance fitter working shifts had a heart attack. He wished to return but was told that he could not because the job involved shift-working, which his doctor had advised him to avoid, and heavy lifting and he was dismissed. The evidence was that he could have done a day job and that the heavy lifting was an excuse – it was minimal and someone else could have done it. The dismissal was held unfair on the facts. However, there are few rigid rules in the law of unfair dismissal because everything depends on what is considered reasonable in the particular instance. Mr McInally worked as a barman at a BP workers' camp at Sullom Voe, in an isolated community. He consulted his own GP who diagnosed that he was suffering from asthenia. Later, he was interviewed by Dr Macaulay, one of BP's OH physicians, who concluded that he had a depressive illness caused by the unusual environmental conditions; he told McInally's employers, a catering firm, that he was medically unfit for that kind of work. The GP agreed. The employee was dismissed and claimed that he had been unfairly treated because he had not been consulted. The Scottish Employment Appeal Tribunal decided that consultation was not necessary in this case because it would have made no difference. There was no job suitable for the employee at Sullom Voe and he had been specifically recruited to work there (*Taylorplan Catering (Scotland)* v. *McInally* (1980)).

Intermittent sickness absence for a variety of different complaints is discussed in 8.20 *infra*.

8.16 Medical Reports

Occupational physicians in general do not give the employer clinical details, but confine themselves to a bald statement that the worker is fit or unfit for work. Where the worker has been off sick for a long period, this is insufficient to enable management to make decisions about the worker's future. In such cases, it is necessary that the physician gives a summary of the employee's medical problem and the likely prognosis *with the employee's consent*.

'There are occasions when the employee is seen subject to a formal process, such as part of a sickness absence or substance abuse policy, and this may be on the instruction of line management or a personnel department. In these circumstances the occupational physician should take particular care to explain precisely the purpose of the assessment, make it clear to the individual that the physician is acting as an impartial medical adviser, and ensure that the employee agrees to the assessment. A misunderstanding is less likely when the purpose of such a referral has been defined in a company policy agreed between management and the workforce.' (Faculty of Occupational Medicine: *Guidance on Ethics for Occupational Physicians* (1993))

The courts have emphasised that the decision to dismiss is a management one. Medical evidence should be available, but it is not for the doctor to decide: he is asked only for an expert opinion. In practice, however, where a doctor advises that an employee is unfit the employer may have little discretion, because continued employment of that person may involve him in legal liability to that employee or to others. Sometimes, the doctors do not agree. In a case in 1980, *Harper* v. *National Coal Board*, an employee, a known epileptic, had worked for a number of years without incident as a dust mask cleaner in a lamp cabin. Then he had three fits within a period of two years during which he displayed violence to other employees. The local OH physician recommended that he be retired on grounds of ill-health, but the Area Medical Officer advised that adjustments in the employee's medication had made further incidents unlikely. The colliery managers preferred the advice of the local doctor and dismissed. The tribunal obviously shared the fears of the managers and upheld the dismissal in a decision which has been criticised for its apparent approval of prejudice:

'Where the belief (that the employee was a danger) is one which is genuinely held, and particularly one which most employers would be expected to adopt, it may be a substantial reason even where modern sophisticated opinion can be adduced to show that it has no scientific foundation.'

In that case the doctors who differed were both specialist OH physicians, but more frequently any disagreement is between an outside doctor and an OH specialist. Such a case was *Jefferies* v. *BP Tanker Co. Ltd* (1979), which concerned a radio officer on an oil tanker who had two attacks of myocardial infarction in one year. BP's Chief Medical Officer advised that he was permanently unfit for duties at sea and there was no shore-based post available, so he was dismissed. A consultant cardiologist paid by the employee's union then advised that there was no reason whatsoever from the medical point of view why the employee should not continue to do the job. An industrial tribunal held the dismissal fair:

'Were they (the employers) to be criticised because they accepted the advice of their own medical officer, who was qualified in occupational medicine ...?'

It is submitted that managers when deciding whether to dismiss on grounds of ill-health are faced with two issues:

(1) What is the state of the employee's health?
(2) Is he fit to do the job he is employed to do or some other job which might be available?

Very few patients will have medical conditions so clear-cut that a doctor can say either that they are definitely fit or definitely unfit. Therefore, he may be obliged (with, of course, the employee's consent) to give some account of the case, for example that the man has had a serious heart attack, has made a good recovery, but that there is a risk of a further attack. He can say from his knowledge of the job that the risk is increased by the long hours, stress and environmental conditions, but in the end it is the manager who, having been made aware of the risk, has to decide if he is willing to take it.

The manager is entitled to take the doctor's opinion as authoritative with regard to the first question, because he is the undoubted expert. Personnel departments are not expected to have medical textbooks on their shelves:

> 'We do not think that an employer, faced with a medical opinion, unless it is plainly erroneous as to the facts in some way, or plainly contains an indication that no proper examination of any sort has taken place, is required to evaluate it as a layman in terms of medical expertise.' (*Liverpool AHA* v. *Edwards* (1977)).

The nature of the employment and possible alternatives are, however, matters where medical opinion, though important, is only one factor to take into account. This is bad news for some managers who prefer to be able to shelter behind a bald statement that the employee 'failed the medical'. Also, there is a growing demand from workers to be given the right of appeal to an independent consultant against the OH physician's report. Increasingly, the OH physician's advice will not be final, because the 'medical' part of it can be appealed to another doctor, and the 'job-related' section can be challenged.

The *Jefferies* case was a decision only of a local tribunal and it therefore has no authority as a binding precedent. A different approach prevailed in *Milk Marketing Board* v. *Grimes* (1986), a decision of the Employment Appeal Tribunal. Mr Grimes was a milk salesman employed by the Milk Marketing Board who had taken over a small company. His duties included driving a milk float. After a month the new employers realised that he was so deaf that the only way he could communicate with others was in writing. The personnel department took the employee off the road until he had had tests and seen their OH doctor, Dr Burgess. The latter concluded that he was unfit for driving duties. A copy of Dr Burgess's report and the audiometry findings were sent to Grimes's GP, who agreed that the deafness was not materially assisted by a hearing aid. Grimes was dismissed with notice. Afterwards, he consulted an ENT surgeon who advised that use of a hearing aid would restore his threshold to a tolerable level and also pointed out that there were many drivers on the road with levels of hearing worse than Grimes. He drew attention to the fact that the employee had a record of 34 years' accident-free motoring in his private car and that he continued to drive it after the dismissal, because there are no regulations, even for HGV or PSV drivers, which prevent a deaf person from driving. Two totally deaf HGV drivers subsequently gave evidence in the industrial tribunal. Grimes also had an accident-free record of several years of driving the milk float. The employers had a disciplinary procedure concerned with dismissals but it only covered dismissals for misconduct, not capability. They stated, however, that they would have allowed him to appeal against his dismissal if he had asked.

The tribunal criticised the employers for not expressly inviting Grimes to obtain his own medical report if he wished. They held that the employee did not have a proper opportunity to challenge Dr Burgess's findings and had therefore been unfairly treated. Two contested issues arose in the Grimes case. One concerned the degree of his deafness and whether it could be improved with a hearing aid, the other whether a profoundly deaf individual is fit to drive on the public highway. Medical evidence must be conclusive on the first, but management should have reviewed the doctor's opinion on the second. The willingness of the tribunal to interfere is probably partly to be explained by the very 'ordinariness' of the employee's job. Industrial tribunals are prepared to accept that an OH physician has special knowledge about working on an oil tanker, but they are more critical of his views on milk float propulsion. There was also the difficulty of communication in Grimes's case which meant that the employers should have taken care to allow him the opportunity to protest.

It is important to identify two separate strands in the case law. One well-established line of decisions requires employers not to act solely on the advice of their medical adviser but to allow the employee to consult his own 'expert witness', in order to put his side of the case. There is, however, little judicial guidance thus far for the manager who receives two conflicting medical reports. Suppose that Grimes procures medical evidence in his favour? Must the employer automatically accept it and reject the report of the occupational health specialist? Considering the disastrous consequences of a poor medical report on the employee's livelihood, it may be thought fair to institute a procedure whereby the employee may be allowed the right to 'appeal' against the occupational physician's opinion to a *third* doctor, a consultant who is independent of both parties. Some large employers now incorporate this final appeal into their procedures, in the event of two conflicting medical reports. As the employer will be paying for this opinion, it is likely that he will wish to nominate the consultant, though it is preferable for the parties to agree on the identity of the 'arbitrator'.

The OH doctor should submit a clear report. In *WM Computer Services Ltd* v. *Passmore* (1987) an accounts controller was off work for some weeks suffering from depressive illness. He attempted suicide and was being treated by a psychiatrist. The employers asked him to see their OH physician, Dr Johnson. The latter wrote both a report and a covering letter. The report stated that the employee was anxious to return, but was unlikely to be able to perform at 'peak level' for some time to come. The letter said that the doctor was 'rather gloomy' in his prognosis, and that the employee was aware that if he did return it might be only for a short period. The doctor admitted in evidence in the tribunal that he was trying 'to sit on the fence'. The employers construed his opinion as an indication that the employee was unfit and decided to dismiss him. They asked Dr Johnson how they should do this and he advised by letter rather than interview. It was held that the employers had acted unfairly. The proper mode of termination was an employment not a medical matter and the employee should have been consulted (presumably he should have been sent a letter inviting him for interview if he wished to come). The employers had 'misconstrued' the doctor's letter, and they should have considered the possibility of alternative employment.

Compare the Scottish case of *Eclipse Blinds* v. *Wright* (1992). Ms Wright was a registered disabled person who had been employed as a receptionist since 1978. Between 1985 and 1987 her health deteriorated and she was frequently off work. After some improvement she became ill again in March 1989, and was absent for several months. The employers contacted her own doctor, with her permission. He indicated that the ultimate prognosis was not good. Ms Wright herself believed that her health was improving and that she would soon return to work. The employers did not wish to engage in a conversation with Ms Wright in which they might have to disclose the gloomy news, so they dismissed her by letter. The Employment Appeal Tribunal upheld the decision of the tribunal that in the exceptional circumstances it was not unfair to fail to consult the employee. Each case rests on its own facts. In the Wright case the employee was unlikely ever to work again, whereas in the Passmore case the tribunal members were of the opinion that there was hope of the employee continuing to hold down a job if he could weather the current crisis. Unemployment is not the best medicine for clinical depression.

The Faculty of Occupational Medicine advises that statements in medical reports such as 'fit for light duties' should be avoided:

'The physician should bear in mind that it is the employer who is responsible for allocating duties even though the decision should take account of constructive professional advice wherever practicable.'

The doctor's job is to assess the capabilities of the worker, not to decide whether he should be dismissed. Sympathy for an employee and the knowledge that the loss of his job may well affect his health must be balanced against the need for the employer to be able to trust in the doctor's objectivity in the interests of the rest of the workforce.

8.17 'No illness' agreements

These stem from a decision of the Scottish courts in *Leonard* v. *Fergus and Haynes* (1979). The employee was a steel fixer who entered into a contract of employment with contractors supplying labour for the construction of a concrete oil platform. Conditions were arduous and it was vital that all the employees were in rude health. A written contract provided that any man absent for two shifts in 14 days would be dismissed as unsuitable for North Sea related work. Leonard was absent for some time because, he alleged, of an industrial injury. The employers dismissed him without consultation and without a medical report; the dismissal was held to be fair. However, employers who put such terms into contracts with all their employees are likely to be disappointed. It was the special nature of the work, not the way the contract was drafted, which led to the decision in favour of the employer. Also, the Court of Appeal has recently regarded with disfavour contracts providing for automatic termination if the employee does not come to work, holding that they are void as an attempt to exclude the protection of the statute (*Igbo* v. *Johnson Matthey Chemicals* (1986)).

8.18 Disabled workers

A common misconception is that the law forbids the dismissal of workers registered as disabled. The Disabled Persons (Employment) Act 1944 obliges employers with 20 or more workers to employ at least 3 per cent registered disabled. It forbids him to discharge a registered disabled person *without reasonable cause* if he is or would thereby fall below quota. 'Reasonable cause' has been interpreted widely. In *Seymour* v. *British Airways* (1983) the employee had an accident at work which left him permanently disabled. He was registered as a disabled person. The employers, being in severe financial difficulties, decided that they would have to make redundant all 'non-effective' staff and dismissed Seymour. The dismissal was upheld as fair. Financial problems constituted reasonable cause for sacking disabled employees, although employers should give consideration to finding alternative work if it was available:

> 'A person who is both disabled and registered as such is entitled to special consideration by his employer. That special consideration includes looking at his personal circumstances before deciding to dismiss him. However, the extent to which a registered disabled person should be given preference over an unregistered or similarly disabled one or over an able-bodied person where redundancies are necessary must be judged by the standard of reasonableness in accordance with the statutory provisions.'

This decision and others have led to a call for special legislation to make discrimination against the disabled unlawful.

Where an employee is taken on by an employer who knows of his disability, it will be unfair to dismiss him because his work is not up to the standard of an able-bodied person. On the other hand, since no-one can complain to a tribunal unless he was employed for at least two years at the date of dismissal, the law gives a two year probationary period.

8.19 Dismissal for misconduct

The ACAS Code of Practice: *Disciplinary Practice and Procedure in Employment* (1977) stresses the need for all employers to have disciplinary rules which should be communicated clearly to all employees. Rules should cover issues like timekeeping, absence (including the need to notify the employer of ill-health and the need for medical certification), health and safety matters (like prohibitions on smoking, drink and drugs), use of company facilities, and the attitude of the employer to racial or sexual abuse. The employer should make it clear which offences are regarded as constituting gross misconduct which could lead to summary dismissal. In one case, because of problems of absenteeism after the staff Christmas party in the previous year, the company and the union agreed the following December that, if as a result of the Christmas party an employee so indulged himself that he was unable to attend for work the next day, he would be summarily dismissed. Mr Skinner, who was dismissed in pursuance of this rule, was successful in a claim for unfair dismissal because the company had not communicated it to the employees in writing or given sufficient warning (*Brooks* v. *Skinner* (1984)).

It is also important to have disciplinary procedures. These should be in writing, specify the levels of management authorised to take disciplinary action, and provide for individuals to be informed of the complaints against them and to be given an opportunity to state their case, accompanied by a trade union representative or fellow employee, before decisions are reached. Except in cases of gross misconduct, no employee should be dismissed for a first breach of discipline. No disciplinary action should be taken until the case has been carefully investigated. Except in very small companies, there should be a right of appeal to a higher level of management.

In the case of minor offences, the individual should be given a formal oral warning, and a note should be kept. In the case of more serious offences or an accumulation of minor offences the employee should be given a formal written warning and, if this is ignored, eventually a final written warning stating that any further infringements will lead to dismissal. Many procedures provide that these warnings will be expunged from the record after a period of satisfactory conduct, but this is for the employer and the union to agree. Only in cases of gross misconduct will this procedure be inapplicable.

The employee who puts himself or others at risk by flouting safety rules is clearly guilty of misconduct. In a serious case this may amount to gross misconduct justifying summary dismissal, in other situations it may be enough to give a warning. In *Martin* v. *Yorkshire Imperial Metals* (1978), the employee was dismissed after it

was discovered that he had tied down with a piece of wire the left
hand lever on the automatic lathe that he operated. This in effect
removed the safety device whereby the machine could be operated
only if the operator used both hands, thus excluding them from the
area of danger. It was held that the dismissal was fair. What of the
worker who is a danger but not deliberately careless? One such
case which reached the Court of Appeal concerned an airline pilot
whose carelessness and incompetence caused a plane crash. He
was held to have been fairly dismissed. His job was too responsible
to risk another mishap (*Taylor* v. *Alidair* (1978)).

Industrial misconduct depends to a large extent on the nature of
the industry. Smoking one cigarette in a mine is a criminal act as
well as justifying summary dismissal, but in an office it may not
even be a disciplinary offence, unless there is a no-smoking rule. A
school teacher who is convicted of a homosexual offence unconnected with his employment will probably lose his job, while a
labourer convicted of the same offence would be unfairly treated if
he were to be dismissed.

8.20   Ill-health and misconduct

It is sometimes difficult to distinguish a case of ill-health from one
of misconduct. To be absent from work through illness is not
misconduct, but failure to turn up for work with no good reason is.
Disciplinary procedures are inappropriate in genuine ill-health
cases, but where the employee has a series of short-term absences
all ostensibly for medical reasons (flu, back-ache, gastritis, nervous
debility and so on) and is never absent for long enough to need a
doctor's note, managers may use procedures analogous to the
disciplinary code. First, they may institute a series of warnings. If
the employee is ill it may be thought illogical to threaten him with
the sack if his health does not improve, because that is presumably
a matter outside his control, but in *International Sports Co* v. *Thomson*
(1980), it was decided that where the employee has an unacceptable
level of absences due to minor ailments (in that case 25 per cent of
working days lost during 18 months) the employer should interview the employee and, if not satisfied, give appropriate warnings
that if there is no improvement in the attendance record the
employee will be dismissed. The justification for this is that the
employer cannot reasonably be expected to continue employing an
unreliable employee, whatever the reason for his unreliability. The
need for warnings reflects, as in the redundancy cases, the philo-

sophy that no-one should lose his job unexpectedly without an opportunity to prepare for the eventuality. If, of course, the manager can prove that the employee is malingering, he can dismiss for misconduct, but in most cases this is impossible to establish.

Further guidance on intermittent sickness absence was given by the Employment Appeal Tribunal in *Lynock* v. *Cereal Packaging* (1988). In determining whether to dismiss an employee with a poor record of intermittent sickness absence, an employer's approach should be based on sympathy, understanding and compassion. A disciplinary approach involving warnings is not appropriate but the employee should be cautioned that the stage has been reached when it has become impossible to continue with the employment. Factors which may prove important include the nature of the illness; the likelihood of it recurring or of some other illness arising; the length of the various absences and the periods of good health between them; the need of the employer to have done the work of the employee; the impact of the absences on those who work with him; the adoption and carrying out of the policy; the emphasis on a personal assessment in the ultimate decision; and the extent to which the difficulty of the situation and the position of the employer have been explained to the employee. There is no principle in such cases that the fact that the employee is fit at the time of dismissal makes the dismissal unfair.

ACAS (*Discipline at Work* (1987)) recommends that where there is no medical advice to support frequent self-certified absences, the employee should be asked to consult a doctor to establish whether medical treatment is necessary and whether the underlying reason for absence is work-related. If the employee is suffering from alcoholism or psychological stress, or diabetes or glandular fever, or other disorders, these may manifest themselves in a series of minor ailments. In all cases the employee should be told what improvement in attendance is expected and warned of the likely consequences if this does not happen. In the case of workers addicted to drugs or alcohol the employer may agree to continue to employ them on condition they submit to treatment, but warn that if the treatment is rejected or proves unsuccessful he will have to consider dismissal. If the unreliable attendance continues, despite warnings, the employer may decide to dismiss. In contrast to cases of dismissal for long-term absence through ill-health, medical evidence is not so vital where there have been frequent absences for minor complaints, because of the variety and fleeting nature of the employee's illnesses.

What if the employee has been told by his doctors that he must take certain precautions which he then proceeds to ignore? A diabetic is instructed that he must monitor his insulin levels and eat at regular intervals, but he is careless. It is submitted that such a case is analogous to a failure to use safety equipment and that it is misconduct justifying dismissal if the employee has been warned that the employer will not tolerate such carelessness, because it mars job performance and carries risk to the diabetic and others.

Company rules frequently provide that anyone found drunk at work will be summarily dismissed, but employers may also have a policy that alcoholics should not be sacked if they agree to treatment. Brown, who for the first time in his life is drunk at work because his wife has had a baby, is dismissed for gross misconduct, but Green, who has been drinking heavily for years, is sent to a nursing home. Sick employees must be treated with understanding, but if they are in a job which carries a risk to others they will have to be suspended from it until they can again be relied upon. Employers, to show consistency and fairness, should have a policy for drink and drugs at work which is made known to the workforce. As with all disciplinary rules, they must be clearly communicated, as must the consequences of a breach.

Though the Health and Safety Executive advocates the setting up of drug policies providing support in confidence for workers who seek help with drug addiction, courts tend to be unsympathetic to those involved with prohibited drugs. One Scottish dental technician who purchased a small amount of cannabis during his lunch break for personal consumption at home was summarily dismissed when he was convicted of a criminal offence. The dismissal was held to be fair (*Mathewson* v. *Wilson Dental Laboratory* (1988)).

8.21   Selection for redundancy

Redundancy in law means that the job, not the worker, has gone. It is caused either by the employer closing down the business where the employee was employed, or by the cessation or diminution of the requirements of the business for employees to carry out work of a particular kind. It is not designed to support the employee while he is unemployed, but to compensate him for the loss of a job, so it is payable even when the redundant employee obtains a better position elsewhere. If the employer genuinely has no further need for the employee's services, it will be fair to make him redundant, but there is a statutory obligation to make a redundancy payment based on the employee's age and length of service to employees who have worked for the employer for at least two years at the date of dismissal. Many employers agree by contract to pay more than the statutory minimum as an incentive to workers to leave. Redundancy money must be paid in full by the employer: the redundancy rebates system to which all employers had to contribute and which was administered by the Department of Employment has been abolished. This has made it easier for employer and employee together to agree a redundancy. An employee who volunteers for redundancy or early retirement will not thereby lose his right to unemployment benefit (Social Security Contributions and Benefits Act 1992). The employer may avoid making a redundancy payment if he offers suitable alternative employment which the employee unreasonably refuses. It is usually reasonable for an employee to refuse a new job which carries less wages, lower status, or involves substantially more travel and inconvenience.

Employees who have been made redundant sometimes take a case to an industrial tribunal to complain that their dismissal is unfair. Often, this is on the ground that they have been selected for redundancy before someone else who is more suitable. The EPCA provides that employers should normally use customary arrangements and agreed procedures (often 'first in last out') in choosing workers to be made redundant. The Employment Appeal Tribunal has held that it is not necessarily unfair to make disabled employees redundant first (*Seymour* v. *British Airways* (1983)). It may be fair to take an employee's attendance record into account in deciding to choose him for redundancy, but only if the reasons for his absence are investigated (*Paine and Moore* v. *Grundy* (1981)). In other cases, the employee objects to the procedure by which he has been dismissed. He says that he was not consulted or given any prior warning, or that he was not considered for alternative employment which he knows was available. All these factors may make a dismissal unfair. Where the employer dismisses employees of a class for which he recognises a trade union, the Trade Union and Labour Relations (Consolidation) Act 1992 directs that he must consult the union about proposed redundancies, otherwise he may be ordered to make a protective award of extra wages to the redundant employees. If the employee establishes that his redundancy was unfair, he will obtain extra compensation in addition to his redundancy payment.

An occupational health department often becomes involved in a case of redundancy or early retirement because these are seen as relatively humane ways of getting rid of an employee whose performance is unsatisfactory because he is unwell. The pension fund rules probably demand evidence that the employee is permanently incapable of doing the job to allow him to take ill-health retirement. 'Redundancy' may be a face-saving formula, but care should be taken to obtain the employee's full agreement, otherwise he may later allege that the redundancy was not genuine and complain of unfair dismissal.

8.22   *Contravention of a statute*

It is a potentially fair reason for dismissal that the employee could not continue to work in his job without contravention (either on his part or on that of the employer) of a duty or restriction imposed by a statute. Thus, it is fair to dismiss a doctor or nurse who has been removed from the register because he can no longer be lawfully employed in a professional capacity. Since the employer is under many statutory duties to protect his employees' health and safety, it may be necessary to dismiss an employee whose continued employment is a risk to himself or to others. Health and safety laws are paternalistic: they protect workers against their own foolishness and disability as well as seeking to protect others. It is a criminal offence under the Health and Safety at Work Act for an employee to disregard his own safety or that of fellow workers. Even employees who are at risk without any fault on their part may have to be removed from danger. In *Yarrow* v. *QIS* (1977) an employee who was employed with unsealed sources of ionising radiation was found to be suffering from psoriasis. His dismissal was held to be fair. The willingness of the employee to run the risk made no difference. Of course, in such cases the employer should try to find alternative work if any is available.

8.23   *Some other substantial reason for dismissal*

This fifth potentially fair reason covers a wide spectrum. If an employer wishes to introduce a change in his employees' terms of employment, such as hours, location or pay structure, for a good business reason and they refuse to agree, their non-cooperation may amount to a fair reason for dismissal, even though the law of contract does not allow him to force the new terms on them against their will (a somewhat hollow reassurance). A refusal by an important customer to accept an employee may justify dismissal, but the employer has to establish that he had no alternative. In *Grootcon* v. *Keld* (1984) a plater on an oil rig owned by BP sustained a knee injury in the course of his work and was sent home three days early. BP sent a telex to Grootcon, his employer, stating that he was not to be allowed to return until cleared by BP's medical officer. Keld was then dismissed. The Employment Appeal Tribunal decided that an ultimatum from BP that the employee should be removed from the rig might have justified his dismissal. On the evidence presented to the tribunal, however, there was no such demand and the dismissal was unfair. Even prejudice within the community will apparently suffice, as in *Saunders* v. *Scottish National Camps* (1980) where the employee, who worked in a children's holiday camp, was dismissed only because he was discovered to be homosexual. And, despite an express provision in the EPCA that a threat by other employees of industrial action if an employee is not dismissed cannot be used by the employer as a defence, it seems that extreme antisocial behaviour (strong body odour, frequent discussion of intimate matters and so on), found objectionable by other workers and not yielding to warnings, can be a fair reason to dismiss.

The Trade Union Reform and Employment Rights Act 1993 creates a new class of unfair dismissal: victimisation. An employee dismissed for bringing proceedings or making a complaint in good faith against the employer to enforce any of his rights under the Employment Protection (Consolidation) Act 1992, the Wages Act 1986 and the trade union rights conferred by the Trade Union and Labour Relations (Consolidation) Act 1992, will automatically be entitled to compensation for unfair dismissal. There is no minimum qualifying period of employment.

8.24 *The importance of procedure*

This section began with a statement that in unfair dismissal it was the reason, not the procedure which was important. However, a fair person does not have a good reason for dismissing an employee unless he has undertaken a reasonable investigation and acts on evidence rather than prejudice. On the whole, procedure is most important in misconduct cases and this is reflected by the creation in 1977 by ACAS of a special advisory Code of Practice for dismissals for misconduct. The House of Lords has held in an important test case that a failure to follow a fair procedure, as dismissing without first hearing the employee's side of the case, refusing to allow him to be accompanied by a trade union official, denying him an internal appeal and so on, is capable of in itself rendering a dismissal unfair, whether or not the employer has a good reason (*Polkey* v. *Dayton Services*) (1987)). Managers should never, therefore, act on the spur of the moment. Even in cases of gross misconduct, the employee should be suspended for a few days and then interviewed with his trade union representative before a decision is reached. If the employee is genuinely redundant he should be given notice of his impending dismissal and the opportunity to make suggestions about suitable alternative employment or other strategies for avoiding redundancy. In an ill-health case, the employee should be consulted and allowed to produce medical evidence, and possible alternative jobs should be considered.

8.25 Strikes and industrial action

Collective action by groups of employees is subject to different rules from those governing the individual employer/employee relationship. Because employers have a duty to take reasonable care of their employees, a worker who is asked to work in unreasonably uncomfortable or dangerous conditions is not acting in breach of contract if he refuses. There may be debate between the parties about the degree of risk. In *Lindsay* v. *Dunlop Ltd* (1980), the employee refused to work in an area where he was exposed to hot rubber fumes after a preliminary report from the HSE that they might be carcinogenic. The concentration of fumes in Dunlop's factory exceeded the threshold limit recommended by the British Rubber Manufacturers' Association. The trade union and the rest of the workforce agreed to use masks as a temporary measure pending a full report; the masks made the job even hotter and more unpleasant than before. Lindsay refused to work, arguing that the employers were under a duty to provide an expensive new ventilation system. The Employment Appeal Tribunal held that it was not their job to investigate whether there had been a breach of the criminal law, but to decide whether the employer had done what was reasonable. The employers had responded to complaints by providing masks and were willing to continue to monitor the situation. It was held that Lindsay had been fairly dismissed.

The legal position would have been quite different if the trade union in similar circumstances had called on all the workers to down tools. An employer who dismisses all those taking strike or other industrial action while the action is in progress and does not reinstate any within three months is immune from proceedings in an industrial tribunal. An employer faced with unofficial action can choose which of the strikers he will punish, but in a case of action officially supported by the union he must dismiss all or none. The rationale for this approach is the reluctance of Parliament to subject the rights and wrongs of industrial conflict to the scrutiny of the courts. The definition of 'industrial action' is much wider than might be supposed. To take a hypothetical example, an employer insists that his employees work with a hazardous substance without any of the advised precautions. The trade union calls on its members to refuse to work until proper protection is provided. This is industrial action however much it is the employer who is in the wrong.

The need to support workers who act in good faith to preserve their health and safety has led to legislation. The Offshore Safety (Protection against Victimisation) Act 1992 protects those acting as safety representatives and members of safety committees on offshore installations. The Trade Union Reform and Employment Rights Act 1993 (TURERA) gives employees the right not to be subjected to any detriment by their employer (including, of course, dismissal) on the ground that:

(1) having been designated by the employer to carry out activities in connection with preventing or reducing risks to the health and safety of employees at work (this would embrace occupational health professionals as well as safety officers and hygienists), he carried out or proposed to carry out any such activities;
(2) he performed or proposed to perform his functions as a safety representative;
(3) he performed or proposed to perform his functions as a member of a safety committee;
(4) being an employee at a place where there was no safety representative or safety committee or it was not reasonably practicable for him to raise the matter through the safety representative or safety committee, the employee brought to the employer's attention, by reasonable means, circumstances connected with his work which he reasonably believed were harmful or potentially harmful to health and safety;
(5) he left or proposed to leave his work in circumstances of danger which was serious and imminent and which he could not reasonably be expected to avert; or

(6) he took or proposed to take appropriate steps to protect himself or other employees from danger in circumstances of serious and imminent danger, but not if those steps were so negligent that the employer acted reasonably in disciplining him.

This leaves a great deal to the judgment of the industrial tribunal to which any complaint must be made. Serious or imminent danger is to be judged from the perspective of the reasonable belief of an employee of normal fortitude. There is no minimum qualifying period of employment. An enhanced special award of compensation is payable to those penalised under (1) and (2) or (3) above.

The employer is prohibited from taking disciplinary action against trade union representatives who complain of unfair or unhealthy conditions. This protection covers safety representatives, under the Safety Representatives and Safety Committee Regulations 1977, and others acting on behalf of the union, like a shop steward, under the Trade Union and Labour Relations (Consolidation) Act 1992. It does not extend to rank and file members acting unofficially on behalf of fellow workers. In *Chant* v. *Aquaboats* (1978) a union member (not a safety representative) complained about safety standards at work and organised a petition of other employees which was vetted by the local union branch. When he was dismissed soon after, he claimed that the real reason was his petition. Chant had not been employed long enough at the date of dismissal to be able to complain of unfair dismissal unless the reason was participation in union activities for which there is no qualifying period of employment. The Employment Appeal Tribunal held that the activities were not those of a trade union but of a trade unionist, so Chant lost his case. The activity must be with union authority rather than as an individual. It follows that a shop steward who acts against union policy is not participating in the activities of the union. Chant might now be protected by TURERA (above) in that in good faith he took reasonable steps to bring a perceived hazard to the employer's attention.

8.26 *Time off for trade union activities*

Special consideration is given by the Trade Union and Labour Relations (Consolidation) Act 1992 to members and officials of unions recognised by the employer. Recognition means that the employer regularly negotiates with the union and allows it to represent employees in grievance and disciplinary matters. An employer is obliged by the Act to allow an official (including shop stewards) of a recognised independent trade union who is in his employment to have reasonable time off *with pay* for the purpose of enabling him

(1)   to carry out
   (a)   any duties of his, as such an official, which are concerned with negotiations with the employer on industrial relations matters such as terms and conditions of employment, or
   (b)   any other duties of his, as such an official, which are concerned with the performance of any functions connected with those matters which the employer has agreed may be performed by the union, or
(2)   to undergo training in aspects of industrial relations which are relevant to the carrying out of those duties and which have been approved by the TUC or by his union.

An ACAS Code of Practice: *Time off for Trade Union Duties and Activities* gives guidance on the operation of the legislation.

Reasonable time off without pay should be given to members of recognised trade unions to attend union meetings, vote in union elections and so on.

Safety representatives are entitled to time off with pay to carry out their duties and to undergo such training as is reasonable. The Health and Safety Commission has issued an Approved Code of Practice as guidance. The approval by the TUC or the union of a training course is not a legal requirement for safety training courses. Thus, an employer who refused his employee time off to attend a course at a college because he provided what, in his opinion, was a perfectly satisfactory course on his own premises, though not approved by the union and lacking any guidance on representing members with a grievance, was not necessarily in breach of the law; it depended on whether the tribunal thought he had acted reasonably (*White* v. *Pressed Steel Fisher* (1980)).

Chapter 9

Equal Opportunities

9.1 The concept of discrimination

Discrimination is only unlawful if it is on grounds of sex, race, colour or ethnic or national origins, or marital status. Religious discrimination is not prohibited (except in N. Ireland), nor is discrimination against the disabled or homosexuals of either sex. Discrimination against Sikhs has been held to be racial rather than religious, against Moslems and Rastafarians religious rather than racial. The legislation protects all workers, male and female, black and white, and is not confined to employees: it embraces contract workers, young persons working on a job training scheme, members of trade unions and partners in a business. The Equal Opportunities Commission (EOC) and the Commission for Racial Equality (CRE) monitor the legislation, issue guidance which may be in the form of Codes of Practice, and assist complainants to take cases to court. They also have power to conduct their own investigations, at the end of which they may issue a non-discrimination notice, ultimately enforceable through the courts. Industrial tribunals have power to hear complaints of discrimination in employment and can award damages, including compensation for injury to feelings. One vital point to grasp about this legislation is that it is not concerned with motives and that it is possible to act unlawfully if the *effect* of what you do is discriminatory despite the manager's good intentions. For this reason, tribunals cannot award damages, only a declaration of the law, for unintentional indirect discrimination.

The Fair Employment (Northern Ireland) Act 1989 prohibits discrimination on religious grounds in Northern Ireland. The Fair Employment Commission has powers considerably wider than those of the EOC or CRE. Quotas may be set, and all employers must monitor the religious composition of the workforce. The Fair Employment Tribunal can impose damages of up to £30,000.

274

The concept of discrimination is a highly technical and complex one. The law prohibits conduct of three kinds:

(1) direct discrimination;
(2) indirect discrimination;
(3) victimisation.

The employer must not discriminate against either job applicants or existing employees. Failure to interview a candidate, rejection of a job application, failure to promote, the provision of adverse working conditions and dismissal on sexist or racial grounds are all unlawful. The Sex Discrimination and Race Relations Acts deal with discrimination in a number of different fields, but this discussion will be confined to employment. There is no qualifying period of employment, nor minimum number of hours. Positive discrimination is just as unlawful as negative discrimination ('We must have a coloured supervisor or else we shall be accused of discrimination'.) The only exception is the provision of training courses for one sex or racial group to try to give them increased opportunity. The employer is vicariously liable for the discriminatory acts of his employees unless he has done that which is reasonably practicable to prevent them.

In a case of direct discrimination, the employer treats an employee worse on grounds of sex or race than he would someone of the opposite sex or a different race. It is not a defence that the employer has a reason for discrimination. The following excuses are not accepted by the law:

'It can be proved that women have more time off than men.'
'My workers/customers would reject a black supervisor.'
'Women haven't got the stamina for this job.'

In *Ministry of Defence* v. *Jeremiah* (1979), a man employed in an ordnance factory protested that the unpleasant job of making colour-bursting shells was confined to male employees because the women found it dirty and were excused. The Court of Appeal held that this was unlawful direct discrimination on grounds of sex. Exceptionally, the employer may discriminate against one sex if the job falls within one of those listed in the statute where sex is a 'genuine occupational qualification'. Examples are actors, models, and welfare and educational jobs where it is more effective for pupils or clients to be cared for by someone of a particular sex. There is also a particular exception for ministers of religion. Men may since 1983 qualify and practice as midwives.

Sexual or racial harassment is a form of direct discrimination.

The European Commission has promulgated a Recommendation and Code of Practice which describes sexual harassment as 'conduct of a sexual nature, or other conduct based on sex affecting the dignity of men and women at work, including superiors and colleagues'. In *Porcelli* v. *Strathclyde Regional Council* (1986) a woman was subjected by male colleagues to suggestive remarks, unwanted physical contact and a display of pictures of half-clothed women. She was awarded compensation against her employer who had failed to take reasonable steps to protect her from this kind of annoyance.

Indirect discrimination is more subtle. It consists of the application of a requirement or condition with which the proportion of members of one sex or race who can comply is considerably smaller than the proportion of members of another sex or race, and which is to the detriment of the former. The following are examples of indirect discrimination:

'Part-time workers will be made redundant before full-time workers.' (Women find it more difficult to work full-time.)

'Applicants must have GCSE English.' (Those whose native language is not English will be less likely to be able to comply.)

'Applicants must be under the age of 35.' (In practice, women find it more difficult to be active in the job market in their 20s and 30s.)

Indirect discrimination is permitted if the employer can show that the requirement or condition is *justifiable*, that is based on some objective job-related requirement unconnected with the sex or race of the worker. Part-time workers may be economically less valuable to the company, GCSE English may be needed for workers whose job requires communication skills and the age profile of the department may be such that younger members are needed (*University of Manchester* v. *Jones* (1993)).

The exclusion of male Sikhs because of their religious requirement to wear a turban has been held justifiable if safety or hygiene are at risk. Statutory regulations give Sikhs exemption from the need to wear hard hats on construction sites, but the employer will not be liable to a Sikh who incurs injury which would have been avoided had he been wearing a hard hat (Construction (Head Protection) Regulations 1989). These regulations do not apply to workplaces other than construction sites: there, the employer will continue to be free to exclude Sikhs without hard hats if that is a necessary safety precaution. It has been held that employers who operate rules prohibiting the wearing of beards by employees in a food factory are justified by hygiene in discriminating against Sikhs (*Panesar* v. *Nestlé* (1980)).

Victimisation is committed when the employer 'punishes' his employee for complaining in good faith about alleged discrimination or supporting someone else's complaint.

An indirect method of enforcing anti-discrimination laws is by what has come to be known as 'contract compliance'. A large organisation may make it a term of its contracts with other organisations that the contractor must respect certain principles, e.g. the rights of his employees to belong to trade unions or a ban on South African imports. The Local Government Act 1988 prohibits certain public, including local, authorities from imposing such terms or withholding contracts from any person for such non-commercial reasons. However, it exceptionally allows the authority to ask approved (by the Secretary of State) questions relating to the racial composition of and opportunities afforded to the contractor's workforce, and to seek undertakings that the contractor is complying with the Race Relations Act. It is also lawful to make contracts obliging the contractor to comply with health and safety legislation.

Breach of the Local Government Act is not a criminal offence, but may give a right to damages. It was strenuously, but unsuccessfully, urged in Parliament that a similar exception be created for contracts specifying the employment of disabled people.

The Sex Discrimination and Race Relations Acts fix an upper limit on the amount of compensation which can be awarded by an industrial tribunal, at present £11,000. This may include a sum for injury to feelings. A highly qualified Asian woman scientist was awarded £3,000 for injury to feelings when her job application was treated with contempt by a health authority appointments committee (*Noone* v. *NW Thames RHA* (1988)). The European Court in *Marshall* v. *Southampton AHA* (No. 2) (1993) decided that the upper limit infringed the European Equal Treatment Directive which ordains that adequate compensation must be awarded by national courts for acts of sex discrimination. The ruling originally applied only to public sector employees, but was extended by statutory instrument to employees in the private sector in November 1993. The European Court also determined that interest must be paid on damages. Since dismissal for pregnancy is an act of sex discrimination if a man off sick for a similar period would not have been dismissed, several women sacked from lucrative jobs when they became pregnant have achieved substantial awards. The upper limit on compensation continues to apply to acts of racial discrimination, where there is no parallel European law.

9.2   *Special laws protecting women workers*

'The Victorian idea of woman was as a wife and mother, centre of the family, consequently the guardian of all Christian and domestic virtues' (Equal Opportunities Commission Report on Health and Safety Legislation (1979)). The conditions in factories in the nineteenth century were abominable for men, women and children, but the plight of women and children most aroused public sympathy. In contrast, the Victorians on the whole did not object to working class women being exploited in domestic service, laundries and dressmaking, which were regarded as their proper place. Male workers resented women working in factories, fearing that women would take jobs from men. The first laws protecting women workers were the Factory Acts 1844 and 1847 which reduced working hours for adult women and forbade the employment of women underground in mines. They were in fact the first victories in the campaign to limit working hours for all employees, both male and female, but they began the industrial segregation of the sexes. Women objected to the restriction of employment opportunity. Some women even dressed as men to work down the pit:

> 'Protective legislation was both a cause and effect of the emergence of the housewife as the dominant mature female role, whether or not a woman worked outside the home. Thus it became possible for employers to employ women on terms that would have been unacceptable to men. It followed that women were effectively excluded from many jobs and skills, and restricted to "women's work".'
>
> (Equal Opportunities Commission (1979))

The hypocrisy of the system became obvious in the twentieth century during both World Wars when restrictions were suspended to allow women to do all kinds of work for which they were supposed to be unfitted by their sex. But women workers themselves by this time mostly supported protective legislation, because they also had to carry the burden of domestic chores. The Equal Opportunities Commission has argued that as long as women have to be treated differently the employer has an excuse for refusing equality of opportunity but, to some extent, the move towards the abolition of these laws giving special treatment to women workers, strongly influenced by the policy-makers in the EC, has been against the wishes of the average British woman on the shop floor.

The new philosophy holds that there are in principle no tasks which a woman is constitutionally unable to perform. Strong women, like strong men, can work down a mine and do heavy manual work. The unique quality of women workers is not that they are frail creatures unable to tolerate harsh physical conditions and bad language, but that they are capable of giving birth. Therefore, special legislation is only justified where it is necessary to protect the mother or the unborn child. All legislation should be examined objectively to determine:

(1) whether it is really needed – for example, do women need the separate provision of lavatories?
(2) whether protection should be extended to *both* sexes – for example against dangerous machinery or inessential night work.

The Sex Discrimination Act 1986 has abolished many restrictions on women's hours of work. The general obligations under the Health and Safety at Work Act which remain state that the employer must do that which is reasonably practicable to ensure the health and safety of all his employees, both male and female, so that it will be unlawful to impose long hours or fail to provide rest breaks if this causes ill-health or accidents caused by fatigue. Statutory rules prohibiting the employment of women on the night shift (subject to the grant of exemption by the Health and Safety Executive) were finally repealed in February 1988.

The Treaty of Rome (the treaty which created the European Economic Community) provided in Article 119 for equal pay for men and women doing equal work. This principle has been developed in a series of directives agreed by the Council of Ministers. The Equal Treatment Directive 1976 provides that:

'... the principle of equal treatment shall mean that there shall be no discrimination whatsoever on grounds of sex either directly or indirectly ... without prejudice to provisions concerning the protection of women, particularly as regards pregnancy and maternity'.

After Miss Marshall, an employee of the Southampton Health Authority, successfully took a case to the European Court, claiming that, by forcing her to retire at 60 when men were allowed to continue to 65, her employer was in breach of this Directive (*Marshall* v. *Southampton AHA* (1986)), the UK Parliament passed the Sex Discrimination Act 1986 making it unlawful for an employer to discriminate in the age of retirement. Four years later, the European Court held in the landmark case of *Barber* v. *Guardian Royal Exchange Assurance Group* (1990) that benefits paid under a contracted-out, private occupational pension scheme must be available to men and women at the same age. This decision was not to be retrospective, but to apply to pensions paid after 17 May 1990. This Delphic ruling was clarified in a protocol to the Maastricht treaty which declares that only service after the relevant date will be affected by the judgment. A differential State retirement age is impossible to maintain in these circumstances, and the UK Government announced late in 1993 that this would become 65 for both men and women workers.

The interpretation of European Community law is ultimately for the Court of Justice of the European Communities in Luxembourg. Conflict between the law of a Member State and European law may occasionally arise. If the European Court decides that European law is directly applicable, the courts of the Member States will have to implement that decision in preference to their own rules. One illustration is the case of Mrs Johnston. She was a policewoman employed in the Royal Ulster Constabulary in Northern Ireland. A policy decision was made by the Chief Constable to arm all male police officers in the regular course of their duties, but that women would not carry firearms. This meant that women officers were unable to perform many routine operations, and Mrs Johnston was told that her contract would not be renewed. UK law allowed the Secretary of State to authorise an act of discrimination by issuing a certificate that it was for the purpose of safeguarding national security and protecting public safety and order, and he had done so in this case. This certificate was by statute to be treated as conclusive. Mrs Johnston, realising that her action in the UK courts was bound to fail, appealed to the Luxembourg court that the Chief Constable's decision contravened European law. They ruled that it was contrary to Community law to try to remove any issue from the jurisdiction of the courts, so that the Secretary of State's certificate should have been reviewed. They also decided that the ban on women carrying guns could only be justified if it could be proved that women were under an inherently greater danger than men because of their sex, which could not be established. A chivalrous desire to protect women as 'the weaker sex' was insufficient (*Johnston* v. *RUC* (1986)).

The Employment Act 1989 repealed all protective legislation, other than that imposed for biological reasons. There is little justification for sex-based discrimination against women working underground in a mine or cleaning machinery in a factory, but there are good medical reasons for special protection for women working with lead or ionising radiations, because of risks to a possible fetus. The Act allows all employers to continue to take special precautions in regard to women of child-bearing potential, whether or not the job is governed by specific regulations, as long as there is scientific evidence of risk. The Sex Discrimination Act 1975 provides that no account shall be taken of special treatment afforded to women in connection with pregnancy or childbirth (though it does not explicitly cover *potential* reproductive capacity). The Employment Act 1989 states that compliance with legislation prior to the Sex Discrimination Acts (like the Health and Safety at Work Act) shall not be an automatic defence to an allegation of unlawful sex discrimination. However, nothing shall render unlawful any act done by a person in relation to a woman if it was *necessary* (not merely advisable) to comply with an existing statutory provision protecting women as regards pregnancy or maternity or other circumstances giving rise to risks specifically affecting women. Certain protective laws are preserved by name in the new provisions. They include regulations limiting the exposure of women workers to lead, regulations excluding pregnant women from work with ionising radiations and regulations excluding pregnant women in some circumstances from working on a sea-going vessel or as flight crew in an aircraft. Restrictions preventing women from being employed underground in a mine or cleaning machinery in a factory have been repealed.

An EC directive on the Protection of Pregnant Women at Work was agreed by the Council of Ministers by a majority in 1992, the UK Government abstaining. The main points of the directive are that:

(1) a specific regime must be established for the review of health and safety risks to pregnant women;
(2) women should not be obliged to work on night work during their pregnancy and for a period following childbirth;
(3) protected workers are entitled to 14 weeks' continuous maternity leave, including at least two weeks' compulsory leave before and after childbirth;
(4) workers are entitled to time off without loss of pay for antenatal examinations;
(5) dismissal of workers during the period beginning with conception and ending with the end of their maternity leave is prohibited, save in exceptional circumstances unconnected with the pregnancy. If a worker is dismissed during this

period the employer must provide written reasons for the dismissal; and

(6) rights under workers' contracts of employment should be preserved during maternity leave. Pay should not be less than she would receive if she were absent for a sickness related reason.

The directive was enacted into UK law in the Trade Union Reform and Employment Rights Act 1993 (see below).

There has already been debate about whether it is necessary to impose controls on the employment of all women, when many are unlikely to be pregnant, through age, sterilisation, or contraception. The Euratom Directive of 1980, revised in 1984, imposes specially low limits on the exposure of 'women of reproductive capacity' to ionising radiation, but it does not define who should be included in the definition. Germany applies the standard to all females under the age of 45, whereas the UK Ionising Radiations Regulations and Approved Code of Practice (1985) give a discretion to the doctor monitoring the individual worker. The employer is obliged to inform women employees engaged in work with ionising radiation of the possible hazard to the fetus in early pregnancy and of the importance of informing the employer as soon as they discover that they have become pregnant. The Control of Lead at Work Regulations 1980 and Approved Code of Practice 1985 also protect pregnant women and those of child-bearing potential, as defined by EMAS or an Appointed Doctor, by excluding them or limiting the permitted exposure.

The Equal Opportunities Commission has pointed out on a number of occasions that levels of exposure to dangerous substances should be reduced for all workers, male and female. There is evidence that men's exposure to hazards may also affect reproductive capacity (Fletcher: *Reproductive Hazards at Work* (1986)). It may not be sufficient to abolish special laws protecting women if both sexes are then exposed to unacceptable risks. Troup and Edwards in their HSE paper *Manual Handling* (1985) conclude that neither sex nor age are reliable predictors or criteria for ability to handle loads. It does not therefore follow that all laws relating to heavy lifting should be abolished, but that employers should be prohibited from requiring any employee, male or female, to handle loads likely to injure. Each employee should be assessed as an individual. This principle is now incorporated in the Manual Handling Regulations 1992 (see Chapter 5). The Code of Practice advises that allowance should be made for pregnancy because it has significant implications for the risk of manual handling injury. Hormonal change can affect the ligaments, increasing the susceptibility to injury; and postural problems may increase as the pregnancy progresses. Particular care should be taken for women who handle loads during the three months following a return to work after childbirth.

Doctors and nurses are apprehensive that they may be caught between Scylla and Charybdis. On the one hand they are told that they may not discriminate against women, on the other that they will be liable to compensate any child born disabled through their negligence in failing to protect it in the womb (Congenital Injuries (Civil Liability) Act 1976 (Chapter 7)). They may be reassured that if they act according to the findings of scientific research and the generally recognised standards of their profession they will steer a safe course. If, however, they either exclude women because of prejudice unsupported by science, or fail to guard them against risks where there is published evidence to show a danger to health, they may come to grief.

The laws against sex discrimination permit the employer to demand strength from a worker if the job requires it, but not to assume that all women are weak and all men like Hercules. A good example was *Shields* v. *Coomes* (1978). A male counterhand in a betting shop was paid more than the women with whom he worked on the assumption that as a man he would bear the responsibility of dealing with any violent disturbance. The employer could only justify unequal pay if he could prove that there was a genuine material difference between the man's job and the women's (Equal Pay Act 1970). Lord Denning said this in deciding that the women were entitled to equal pay:

> 'It would be otherwise if the difference was based on any special personal qualification that he had; as, for instance, if he was a fierce and formidable figure, trained to tackle intruders ... But no such special personal qualification is suggested ... He may have been a small and nervous man, who could not say boo to a goose. She may have been as fierce and formidable as a battle-axe.'

The Equal Pay Act and Article 119 of the Treaty of Rome allow an employee to claim equal pay with a comparator of the opposite sex who is employed by the same employer at the same establishment if he or she can show that he or she is doing the same job or a job of equal value. An industrial tribunal can be asked to appoint an independent job evaluation expert to decide whether jobs are of equal value. The employer may defend the claim if he can show a material difference other than sex between the man and woman, like additional qualifications, seniority, or the need to attract employees with particular skills in short supply.

If the employer sets up his own job evaluation exercise, he will not be permitted to weight it unfairly in favour of either sex by artificially giving more points for manual strength or dexterity. In *Rummler* v. *Dato-Druck* (1987), a woman printer argued that she should be placed in a higher pay grade because the work was, for her, heavy physical work. The same work, she said, did not require so much effort from a man. The European Court rightly held that it would be discriminatory to use values which differed from sex to sex. Women cannot be given more points for doing heavy manual work than men doing the same work. Nevertheless, it is difficult to design a job evaluation scheme which is totally objective, since it will always to some extent reflect the values of the designer. If society undervalues caring skills, seeing them as a natural attribute of the female and not requiring any particular effort, it may not give a high score to a job like nurse or home help.

9.3   *Pregnancy dismissals and maternity leave*

The Employment Protection (Consolidation) Act (EPCA) provided in 1978 that dismissal for pregnancy or a pregnancy-related illness was automatically unfair. For a woman to claim the benefit of this provision, she had to have worked full-time for at least two years at the date of dismissal. Later cases managed to avoid the need to establish this qualifying period of service by showing that dismissal for pregnancy is a form of sex discrimination, and holding an employer liable if it could be proved that he would not have dismissed a man who was off sick for a lengthy period (there is no minimum period of employment under the Sex Discrimination Act).

The UK has now been forced to expand protection for pregnant employees in line with the EC Pregnancy Directive (see above). The new laws are to be found in the Trade Union Reform and Employment Rights Act 1993 (TURERA). Dismissal of an employee is automatically unfair if the reason or principal reason for the dismissal is:

(1)   that she is pregnant or any reason connected with the pregnancy (e.g. miscarriage, hypertension);
(2)   that she has given birth to a child or any reason connected

with the birth, and that she was dismissed during the *maternity leave period*;

(3)   that she took or availed herself of the benefits of, maternity leave;

(4)   where before the end of her maternity leave period she gave the employer a medical certificate stating that she would be incapable of returning to work after the end of the period, and she was dismissed within four weeks of the end of the maternity period;

(5)   that she is subject to medical suspension from her job on maternity grounds;

(6)   that she is made redundant during her maternity leave period and has not been offered suitable alternative employment.

If an employee's pregnancy, childbirth or breast-feeding makes it inadvisable on medical grounds that she should continue with her work a new right to medical suspension has been created. This will only come into play where there is a statutory requirement (e.g. work with ionising radiations) or recommendation in an Approved Code of Practice (e.g. manual handling) under the Health and Safety at Work Act. Where this is the case, the employee must be given any suitable available alternative work or, if there is none, her normal pay during the period of suspension.

A pregnant employee has the right under the EPCA to be given reasonable time off work to attend ante-natal care recommended by a doctor, midwife or health visitor. She is entitled to time off with pay. After the first visit, the employee may be required to produce a certificate that she is pregnant and a record of the appointments made for her.

The identification of a pregnancy-related reason for dismissal has caused problems of interpretation. Two examples of cases decided under the Sex Discrimination Act (not the EPCA because the women had not worked for two years at the date of dismissal) provide an illustration. In *Webb* v. *EMO* (1993) an employer advertised for a temporary substitute for an employee who was about to go off on maternity leave. The substitute discovered soon after beginning work that she too was pregnant and was dismissed. The House of Lords held that the reason for the dismissal was not the pregnancy so much as the inability of the substitute employee to fulfil her contract. A man who announced that he would have to be absent for several weeks in similar circumstances would have been treated the same. This has been referred to the European Court because it arguably conflicts with decisions of that court. In

Hopkins v. *Shepherd* (1994) a trainee veterinary nurse was dismissed when she announced her pregnancy because of risks to the baby from X-rays and infections. The dismissal was held to be fair. It is difficult to see how the tribunal could avoid holding both these dismissals unfair under the new legislation, but the judgment of the European Court in *Webb* v. *EMO* may give guidance.

Central to the new provisions is the concept of maternity leave. An employee has a right to 14 weeks' maternity leave during which all her contractual rights must be preserved as though she had not been absent, except that the employer does not have to pay her her normal wages unless obliged to do so by her contract of employment. Maternity leave starts on the date which the employee notifies to the employer at least 21 days before that date or as soon as is reasonably practicable. If she is absent from work for a pregnancy related reason at any time after the beginning of the sixth week before the birth the maternity leave is triggered by the first day of absence. The Factories Act prohibits the employment of women in factories within four weeks of childbirth. It is likely that new health and safety legislation will extend this (or at least two weeks as required by the Pregnancy Directive) to all employees, and this may prolong the maternity leave. If the child is born prematurely the leave begins on the day of its birth.

The rights of pregnant employees who have worked full-time for the employer for two years at the eleventh week before the baby is due are more extensive. They have, in effect, 40 weeks' leave, because the employer must allow them to return to work for up to 29 weeks after the birth. The employee who resigns before the eleventh week surrenders this entitlement. There is no law which compels the employee to stop working at that stage. In *ILEA* v. *Nash* (1979), the employee was a school teacher whose baby was due at the end of September. She chose to resign with effect from the beginning of September, realising that she would be on holiday during August. It was held that the employers could not force her to leave at the end of the summer term: the decision was hers.

There is a complicated scheme of notification of intention to return which is regularly criticised by the courts because it is so easy for the employee to lose her rights by becoming entangled in red tape. She must give the employer written notification before she goes on maternity leave and again, if he asks, after the birth. The employer may write to her after seven weeks after the expected week of confinement (which she will have stated in her original notice), requesting a written notification that she still intends to return. The letter must tell her that if she does not reply within fourteen days (or if she is ill or on holiday as soon as reasonably practicable) she will lose her right to return. A woman who is in doubt is advised to give notice of intention to return, for there are no penalties if she changes her mind. It has, therefore, become common for employers to agree to give employees on maternity leave more than they are required by law to pay only on condition that the employee returns to work for at least a few months.

The employee must return to work by 29 weeks after the baby is born, unless she can produce a doctor's certificate that she is ill. This postponement can prolong her leave only for up to four weeks, unless her employer agrees voluntarily to hold the post open for a longer period. The employer may also postpone her return for up to four weeks. She must normally be restored to a post in the same grade and capacity as before, unless the employer shows that this is not reasonably practicable and offers her suitable alternative work which she unreasonably refuses. Employers with five or fewer employees are not obliged to allow her to return if it is not reasonably practicable because there is no work for her to do. Otherwise, failure to reinstate as required by the statute is unfair dismissal. As with all statutory rights, the employer can agree better terms with his employees (many employers give maternity leave after only a year's employment), but he cannot fall below the statutory minimum. Temporary replacements may be fairly dismissed when the permanent employee returns as long as they were initially notified in writing that this would happen.

The right to return under the EPCA is a right to return to full-time working: the employer is not obliged to allow the new mother to work part-time. However, in one case, *Home Office* v. *Holmes* (1984), a woman successfully used the sex discrimination laws to establish that a requirement to work full-time is one which discriminates indirectly against women because a smaller proportion of them can work full-time because of family commitments. The Home Office was unable to justify its insistence that she work full-time after the birth of her second child. The Kemp-Jones Report had stated that the civil service was losing valuable trained personnel when they left to start families and that in some departments efficiency increased with the introduction of part-timers. In *Greater Glasgow Health Board* v. *Carey* (1987), in contrast, the Employment Appeal Tribunal decided that a requirement that a health visitor work a five-day week, though discriminatory, was justifiable by the need for continuity of attendance on mothers on a daily basis. What happened at weekends?

9.4   *Maternity pay*

Employees are entitled in some circumstances to Statutory Maternity Pay (SMP) under a system which came into force in 1987. The qualifying conditions are that the employee must have been employed for at least 26 weeks up to and into the fifteenth week before the expected week of confinement, that she must normally give 21 days' notice of her intention to stop work, that she must have been earning on average more than the NI lower earnings limit, that she must provide medical evidence of pregnancy or birth and that she must have stopped work. The right to SMP does not depend on the woman's intention to return to work. Self-employed or non-employed women do not qualify for SMP but can claim maternity allowance from the DSS. There are two rates of SMP, the higher rate which is payable for the first six weeks only, and the lower rate. The employee must have worked full-time for the employer for at least two years by the fifteenth week before the baby is due to receive the higher rate (90 per cent of average weekly earnings). SMP is paid for a maximum of 18 weeks. It is subject to income tax. The employer can claim reimbursement in full of SMP by deducting the money from his National Insurance contributions. Many employers have negotiated better contractual provision, as with sick pay.

The extension of maternity leave to all employees has necessitated changes in the maternity pay rules. The Pregnancy Directive states that during maternity leave pay should not be less than the employee would receive if she were off sick.

The Department of Social Security has issued a consultation document setting out options for implementing the directive. It is accepted that the lower rate of Statutory Maternity Pay will have to be raised to the higher rate of Statutory Sick Pay. The qualifying period will remain at six months. The current rule that reduces a woman's SMP period if she starts her maternity leave later than the sixth week before the expected week of confinement will be removed. Where the woman is absent from work for a health reason unconnected with the pregnancy, she will be able to claim SSP for as long as that illness continues, up to the date of the baby's birth.

Women who have worked for between six months and two years will receive SMP for 18 weeks, as before. Most controversial is the proposal that employers should for the first time contribute up to 8 per cent to SMP (with relief for small employers).

9.5   Parental leave

So far, this is dependent on a contractual agreement with the employer. Movements within the EC towards statutory unpaid parental leave of up to three months for either parent while the child is under two have so far proved unacceptable to the UK government. A short period of leave for the father at the time of the birth of the child is more likely to achieve acceptance here.

9.6   Children and young persons

It was the plight of young children in the factories which inspired the first protective laws at the beginning of the nineteenth century. Soon after, similar legislation was passed to protect children in mines. Gradually, an intricate patchwork of statutes and regulations developed. That was in an era when children became adult wage-earners at a very early age after only rudimentary schooling. When children left school at sixteen and achieved full adult status at eighteen much of the old law had become obsolete. There remains a need for legislation to protect those under the school-leaving age who engage in part-time jobs. No person under 13 may be lawfully employed in any capacity, and from 13–16 only outside school hours and not for more than two hours a day. These restrictions do not apply to approved work experience for children in their last year of school (Education (Work Experience) Act 1973).

There is evidence that people under 25 are more accident-prone than older citizens, but none to show that under-18s are more at risk than the 18–25s. Young people's stamina in the face of long hours, shift work or night work is no less than that of their elders. However, where a young person is not yet fully developed physically, a few occupations may be especially hazardous for teenagers, and the untrained and inexperienced are always at increased risk. Comprehensive measures have been enacted to ensure the health and safety of all employees. The employer owes a duty to do that which is reasonably practicable to protect them, as by giving instruction and training, providing a safe working environment, and fixing hours and meal-breaks in order to avoid risks to health (Health and Safety at Work Act, Chapter 5).

Most of the out-dated restrictions on young people's employment and hours of work were repealed by the Employment Act 1989. A few necessary restrictions survive. Persons under 18 may not operate a power press or circular saw, may not clean machinery in motion, may not work with lead or on an offshore installation, and may not be employed in a bar or gaming establishment.

Many young people at work will be trainees. Often they will be employed while in training, as are apprentices. Where they are not employees but sent to the employer under a youth training, employment rehabilitation, community programme or similar scheme, the protection of the Health and Safety at Work Act is extended to them (Health and Safety (Training for Employment) Regulations 1990).

Under s. 119 of the Factories Act, if a Health and Safety Executive inspector considers that a young person's work in a factory puts his own or others' health at risk, he can stop the young person's employment until he has been medically examined and certified fit. However, s. 10A of the Act makes similar provision for all workers, empowering an Employment Medical Adviser to serve notice on an employer requiring a medical examination of any person or persons if there is reason to believe that that person's health has been or will be injured by the work they are required to do. The Mines Medical Examination Regulations 1964 and the Merchant Shipping (International Labour Conventions) Act 1925 require the medical examination of young persons working underground or at sea. Section 60 of the Health and Safety at Work Act 1974 obliges the health authorities to give the Employment Medical Advisory Service on request details of the school medical record of a young person.

Bibliography

Advisory, Conciliation and Arbitration Service (ACAS) (1977) *Code of Practice 1: Disciplinary Practice and Procedures in Employment.* London.

ACAS (1985) *Advisory Booklet 5: Absence.* London.

ACAS (1987) *Advisory Handbook on Discipline at Work.* London.

ACAS (1991) *Code of Practice 3: Time off for Trade Union Duties and Activities.* London.

American Occupational Medical Association (1986) Drug Screening in the Workplace: Ethical Guidelines. *J. Occup. Med.* **28** 1240–41.

Association of the British Pharmaceutical Industry (ABPI) (1988) *Guidelines for Medical Experiments on Non-Patient Volunteers,* 2nd edn. London.

Atiyah, P.S. (1987) *Accidents, Compensation and the Law,* (ed P. Cane), 4th edn. Weidenfeld and Nicolson, London.

Barrett, B. & James, P. (1988) Safe Systems: Past, Present and Future. *Industrial Law Journal,* **17,** 26–40.

Beveridge, W. (1942) *Social Services and Allied Services.* Cmnd 6404. HMSO, London.

Brazier, M. (1992) *Medicine, Patients and the Law,* 2nd edn. Penguin Books, London.

British Medical Association (BMA) (1984) *The Occupational Physician,* 3rd edn. London.

BMA (1988) *Rights and Responsibilities of Doctors.* London.

BMA (1993) *Medical Ethics Today.* London.

Bunt, Karen (1993) *Occupational Health Provision at Work.* HMSO, London.

Calabresi, G. (1970) *The Costs of Accidents.* New Haven, Connecticut.

Campbell, S. (1986) *Labour Inspection in the European Community.* HMSO, London.

Chamberlain (ed) (1984) *Pregnant Women at Work.* Royal Society of Medicine and Macmillan Press, London.

Commission of the European Communities (1987) *Protective Legis-*

lation for Women in the Member States of the European Community.
COM (87) 105. Luxembourg.

Committee of Inquiry into Health and Safety at Work (Robens Committee)
Report (1972). Cmnd 5034. HMSO, London.

Committee of Inquiry into Industrial Health Services (Dale Committee)
Report (1951). Cmnd 8170. HMSO, London.

Dawson, S., *et al.* (1988) *Safety at Work: The Limits of Self-Regulation.*
Cambridge University Press, Cambridge.

Department of Health (1992) *The Health of the Nation* Cm 1986.
HMSO, London.

Department of Health and Social Security (DHSS) (1984) *Report of*
the Working Party on Confidentiality (Körner Report). London.

Department of Social Security (1986) *Industrial Injuries Handbook for*
Adjudicating Medical Authorities. HMSO, London.

Department of Trade and Industry (1993) *Review of implementation*
and enforcement of EC law in the UK. London.

Dewis, M. (1993) *Health and Safety at Work Handbook.* Tolley
Publications, Croydon.

Discher, Kleinman & Foster (1975) *Pilot Developments of an Occu-*
pational Disease Surveillance Method. University of Washington.

Doll, R. & Peto, R. (1982) *The Causes of Cancer.* Oxford University
Press, Oxford.

Dorward, Anna (1993) *Managers' Perceptions of the Role and Con-*
tinuing Education Needs of Occupational Health Nurses. HSE Books,
London.

Drake, C.D. & Wright, F.B. (1983) *Law of Health and Safety at Work:*
The New Approach. Sweet and Maxwell, London.

Employment Service (1993) *Code of Good Practice on the Employment*
of Disabled People (revised). HMSO, Sheffield.

Equal Opportunities Commission (EOC) (1979) *Health and Safety*
Legislation: Should We Distinguish Between Men and Women?
Manchester.

Fletcher, A.C. (1986) *Reproductive Hazards of Work.* EOC and
Association of Scientific, Technical and Managerial Staffs,
Manchester.

Gardner, W. (1983) *How to Prevent Absenteeism.* Applecross Books,
Southampton.

General Medical Council (GMC) (1993) *Professional Conduct: Fitness*
to Practise. London.

General, Municipal, Boilermakers and Allied Trades Union
(GMBATU) (1987) *Hazards of Work.* Esher, Surrey.

Goddard, G. (1988) Occupational Accident Statistics 1981–85.
Employment Gazette. **96**, 15–21.

Goodman, M.J. (ed) *Encyclopaedia of Health and Safety at Work: Law and Practice.* Sweet and Maxwell, London.

Ham, C., Dingwall, R., Fenn, P., Harris, D. (1988) *Medical Negligence: Compensation and Accountability.* King's Fund Institute, London.

Harvey, S. (1988) *Just an Occupational Hazard?* King's Fund Institute, London.

Health and Safety Commission (HSC) (1978) *Occupational Health Services: The Way Ahead.* HMSO, London.

HSC (1984) *Consultative Document: Control of Substances Hazardous to Health.* HMSO, London.

HSC (1985) *Plan of Work for 1985–86 Onwards.* HMSO, London.

HSC (1986) *The International Labour Organisation Convention 161 and Recommendation 171 on Occupational Health Services.* HMSO, London.

HSC *Plan of Work for 1992/3 and Beyond.* HMSO, London.

HSC *Annual Reports 1990/1, 1991/2 and 1992/3.* HMSO, London.

Health and Safety Executive (HSE) (1982) *Guidance Note MS 18: Health Surveillance by Routine Procedures.* HMSO, London.

HSE (1982) *Guidance Note MS 20: Pre-Employment Health Screening.* HMSO, London.

HSE (1982) *Guidelines for Occupational Health Services.* HMSO, London.

HSE (1984) *Annual Report for the Manufacturing and Service Industries (1983).* HMSO, London.

HSE (1985) *Health at Work, 1983–85. Medical Division of the Employment Medical Advisory Service.* HMSO, London.

HSE (1988) *Review Your Occupational Health Needs: An Employer's Guide.* HMSO, London.

HSE (1990) *Surveillance of People Exposed to Health Risks at Work.* HMSO, London.

HSE (1991) *Workplace Health and Safety in Europe.* HMSO, London.

HSE (1992) *Your Patients and Their Work (for family practitioners).* HMSO, London.

HSE (1993) *Reporting from RIDDOR.* HMSO, London.

HSE (1993) *The Costs of Accidents at Work.* HMSO, London.

House of Lords Select Committee on Science and Technology (1983) *Occupational Health and Hygiene Services (Gregson Report).* HLP 28. HMSO, London.

House of Lords Select Committee on Science and Technology (1984) *Occupational Health and Hygiene Services: the Government Response.* HLP 289. HMSO, London.

Hutter, B.M. & Lloyd-Bostock, S. (1990) The Power of Accidents: the social and psychological impact of accidents and the enforcement of safety regulations *British Journal of Criminology* **30**, 409.

Industrial Injuries Advisory Council (1981) *Industrial Disease: A review of the Schedule and the Question of Individual Proof.* Cmnd 8393. HMSO, London.

Institution of Professionals, Managers and Specialists (1992) *The Third Alternative Report of the Work of the Health and Safety Executive.* London.

International Commission on Occupational Health (ICOH) (1992) International Code of Ethics for Occupational Health Professionals *Bull. Med. Eth* **82**, 7 October 1992.

International Labour Office (1987) *World Labour Report 1–2.* Oxford University Press, Oxford.

James, Phil (1992) Reforming British Health and Safety Law: A Framework for Discussion. *Industrial Law Journal* **21**, 83.

Kahn-Freund, O. (1983) *Labour and the Law*, (eds. P. Davies & M. Freedland), 3rd edn. Stevens, London.

Kelman, G.R. (1985) The Pre-Employment Medical Examination. *Lancet*, 1231–33.

Kloss, D.M. (1988) Demarcation in Medical Practice: the Extended Role of the Nurse. *Professional Negligence* **4**, 41–7.

Lee, W.R. (1973) An Anatomy of Occupational Medicine. *Brit. J. Industr. Medicine* **30**, 111–17.

Lee, W.R. (1973) Emergence of Occupational Medicine in Victorian Times. *Brit. J. Industr. Med.* **30** 118–24.

Lewis, R. (1987) *Compensation for Industrial Injury.* Professional Books, Abingdon.

McCloy, E.C. (1987) Chapter 27. Reproduction and Work. In: *Hunter's Diseases of Occupations.* Hodder and Stoughton, London.

Medical Services Review Committee (1962) *A Review of the Medical Services in Great Britain (Porritt Report).* Social Assay, London.

Morgenstern, F. (1982) *Deterrence and Compensation.* International Labour Organisation, Geneva.

Murray, T.H. (1983) Genetic Screening in the Workplace: Ethical Issues, *J. Occup. Med.* **25**, 451–4.

National Institute for Occupational Safety and Health (1983) *Morbidity and Mortality Weekly Report* (January 21, 1983). Atlanta, Georgia.

Ogus, A.I. & Barendt, E.M. (1993) *The Law of Social Security*, 4th edn. Butterworths, London.

Royal College of Nursing (1987) *Code of Professional Practice in Occupational Health Nursing.* London.

RCN (1991) *Guide to Occupational Health Nursing.* London.

Royal College of Physicians (1989) *Guidelines on the Practice of Ethics Committees in Medical Research*, 2nd edn, London.

Royal College of Physicians (1986) *Research on Healthy Volunteers*. London.

Royal College of Physicians (Faculty of Occupational Medicine) (1993) *Guidance on Ethics for Occupational Physicians*, 4th edn. London.

Royal Commission on Civil Liability and Compensation for Personal Injury (Pearson Commission) (1978). Cmnd 7054. HMSO, London.

Samuels, S.W. (1986) Medical Surveillance: Biological, Social, and Ethical Parameters, *J. Occup. Med.* **28**, 572–7.

Schilling, R.S.F. (ed) (1981) *Occupational Health Practice*, 2nd edn. Butterworths, London.

Silverstone, R. & Williams, A. *The Role and Educational Needs of the Occupational Health Nurse*. Royal College of Nursing, London.

Stapleton, J. (1986) *Disease and the Compensation Debate*. Clarendon Press, Oxford.

Street, H. (1993) *The Law of Torts* (ed. M. Brazier), 9th edn. Butterworths, London.

Troup, J.D.G. & Edwards, F.C. (1985) *Manual Handling: A Review Paper*. Health and Safety Executive. HMSO, London.

United Kingdom Central Council for Nursing, Midwifery and Health Visiting (UKCC) (1992) *Code of Professional Conduct for the Nurse, Midwife and Health Visitor*. London.

UKCC (1992) *Advisory Paper on the Administration of Medicines*. London.

UKCC (1987) *Advisory Paper on Confidentiality*. London.

Wilson, G.K. (1985) *The Politics of Safety and Health*. Clarendon Press, Oxford.

World Medical Association (1983) *Declaration of Helsinki (International Guidelines for Biochemical Research Involving Human Subjects)*. Geneva.

Appendix A

Reporting of Injuries, Diseases and Dangerous Occurrences Regulations 1985

Schedule 2 Reportable diseases

Column 1	Column 2

Poisonings

1 Poisoning by any of the following: Any activity.

(a) Acrylamide monomer;
(b) Arsenic or one of its compounds;
(c) Benzene or a homologue of benzene;
(d) Beryllium or one of its compounds;
(e) Cadmium or one of its compounds;
(f) Carbon disulphide;
(g) Diethylene dioxide (dioxan);
(h) Ethylene oxide;
(i) Lead or one of its compounds;
(j) Manganese or one of its compounds;
(k) Mercury or one of its compounds;
(l) Methyl bromide;
(m) Nitrochlorobenzene, or a nitro- or amino- or chloro-derivative of benzene or of a homologue of benzene;
(n) Oxides of nitrogen;
(o) Phosphorus or one of its compounds.

Column 1	Column 2

Skin diseases

2	Chrome ulceration of:	Work involving exposure to
(a)	the nose or throat; or	chromic acid or to any other
(b)	the skin of the hands or forearm.	chromium compound.

3	Folliculitis.	Work involving exposure to mineral
4	Acne.	oil, tar, pitch or arsenic.
5	Skin cancer.	

| 6 | Inflammation, ulceration or malignant disease of the skin. | Work with ionising radiation. |

Lung diseases

| 7 | Occupational asthma. | Work involving exposure to any of the following agents— |

(a) isocyanates;

(b) platinum salts;

(c) fumes or dusts arising from the manufacture, transport or use of hardening agents (including epoxy resin curing agents) based on phthalic anhydride, tetra-chlorophthalic anhydride, trimellitic anhydride or triethyl-enetetramine;

(d) fumes arising from the use of rosin as a soldering flux;

(e) proteolytic enzymes;

(f) animals or insects used for the purposes of research or education or in laboratories;

(g) dusts arising from the sowing, cultivation, harvesting, drying, handling, milling, transport or storage of barley, oats, rye, wheat or maize, or the handling, milling, transport or storage of meal or flour made therefrom.

Column 1	Column 2
8 Extrinsic alveolitis (including Farmer's lung).	Exposure to moulds or fungal spores or heterologous proteins during work in— (*a*) agriculture, horticulture, forestry, cultivation of edible fungi or malt-working; or (*b*) loading or unloading or handling in storage mouldy vegetable matter or edible fungi; or (*c*) caring for or handling birds; or (*d*) handling bagasse.
9 Pneumoconiosis (excluding asbestosis).	1. (*a*) The mining, quarrying or working of silica rock or the working of dried quartzose sand or any dry deposit or dry residue of silica or any dry admixture containing such materials (including any activity in which any of the aforesaid operations are carried out incidentally to the mining or quarrying of other minerals or to the manufacture of articles containing crushed or ground silica rock); (*b*) the handling of any of the materials specified in the foregoing sub-paragraph in or incidental to any of the operations mentioned therein, or substantial exposure to the dust arising from such operations. 2. The breaking, crushing or grinding of flint or the working or handling of broken, crushed or ground flint or materials

Column 1	Column 2
Pneumoconiosis – *continued*	containing such flint, or substantial exposure to the dust arising from any of such operations.

<table>
<tr><td></td><td></td><td colspan="2">3.   Sand blasting by means of compressed air with the use of quartzose sand or crushed silica rock or flint, or substantial exposure to the dust arising from such sand blasting.</td></tr>
</table>

3.   Sand blasting by means of compressed air with the use of quartzose sand or crushed silica rock or flint, or substantial exposure to the dust arising from such sand blasting.

4.   Work in a foundry or the performance of, or substantial exposure to the dust arising from, any of the following operations:

   (*a*)   the freeing of steel castings from adherent siliceous substance;

   (*b*)   the freeing of metal castings from adherent siliceous substance:

      (i)   by blasting with an abrasive propelled by compressed air, by steam or by a wheel; or

      (ii)   by the use of power-driven tools.

5.   The manufacture of china or earthenware (including sanitary earthenware, electrical earthenware and earthenware tiles), and any activity involving substantial exposure to the dust arising therefrom.

6.   The grinding of mineral graphite, or substantial exposure to the dust arising from such grinding.

7.   The dressing of granite or any igneous rock by masons or the crushing of such materials, or

Column 1	Column 2

Pneumoconiosis – *continued*

    substantial exposure to the dust arising from such operations.

8.   The use, or preparation for use, of a grind-stone, or substantial exposure to the dust arising therefrom.

9.   (a) Work underground in any mine in which one of the objects of the mining operations is the getting of any mineral;
(b) the working or handling above ground at any coal or tin mine of any minerals extracted therefrom, or any operation incidental thereto;
(c) the trimming of coal in any ship, barge, or lighter, or in any dock or harbour or at any wharf or quay;
(d) the sawing, splitting or dressing of slate, or any operation incidental thereto.

10. The manufacture, or work incidental to the manufacture, or carbon electrodes by an industrial undertaking for use in the electrolytic extraction of aluminium from aluminium oxide, and any activity involving substantial exposure to the dust arising therefrom.

11. Boiler scaling or substantial exposure to the dust arising therefrom.

10   Byssinosis.

Work in any room where any process up to and including the weaving process is performed in a factory in which the spinning or manipulation of raw or waste cotton or of flax, or the weaving of cotton or flax, is carried on.

Column 1	Column 2
11 Mesothelioma. 12 Lung cancer. 13 Asbestosis.	(*a*) The working or handling of asbestos or any admixture of asbestos; (*b*) the manufacture or repair of asbestos textiles or other articles containing or composed of asbestos; (*c*) the cleaning of any machinery or plant used in any of the foregoing operations and of any chambers, fixtures and appliances for the collection of asbestos dust; (*d*) substantial exposure to the dust arising from any of the foregoing operations.
14 Cancer of a bronchus or lung.	Work in a factory where nickel is produced by decomposition of a gaseous nickel compound which necessitates working in or about a building or buildings where that process or any other industrial process ancillary or incidental thereto is carried on.

Infections

15 Leptospirosis.	Handling animals, or work in places which are, or may be infested by rats.
16 Hepatitis.	Work involving exposure to human blood products or body secretions and excretions.
17 Tuberculosis.	Work with persons or animals or with human or animal remains or with any other material which might be a source of infection.
18 Any illness caused by a pathogen referred to in column 2, opposite.	Work involving a pathogen which presents a hazard to human health.
19 Anthrax.	Any activity.

Column 1	Column 2

Other conditions

20 Malignant disease of the bones.
21 Blood dyscrasia.

Work with ionising radiation.

22 Cataract.

Work involving exposure to electro-magnetic radiation (including radiant heat).

23 Decompression sickness.
24 Barotrauma.

Breathing gases at increased pressure.

25 Cancer of the nasal cavity or associated air sinuses.

1. (*a*) Work in or about a building where wooden furniture is manufactured;
(*b*) work in a building used for the manufacture of footwear or components of footwear made wholly or partly of leather or fibre board; or
(*c*) work at a place used wholly or mainly for the repair of footwear made wholly or partly of leather or fibre board.
2. Work in a factory where nickel is produced by decomposition of a gaseous nickel compound which necessitates working in or about a building or buildings where that process or any other industrial process ancillary or incidental thereto is carried on.

26 Angiosarcoma of the liver.

(*a*) Work in or about machinery or apparatus used for the polymerization of vinyl chloride monomer, a process which, for the purposes of this provision, comprises all

Column 1	Column 2
Angiosarcoma –*continued*	operations up to and including the drying of the slurry produced by the polymerization and the packaging of the dried product; or
	(*b*) work in a building or structure in which any part of that process takes place.
27 Cancer of the urinary tract.	Work involving exposure to any of the following substances—
	(*a*) alpha-naphthylamine, beta-naphthylamine or methylene-bis-orthochloroaniline;
	(*b*) diphenyl substituted by at least one nitro or primary amino group or by at least one nitro and primary amino group (including benzidine);
	(*c*) any of the substances mentioned in sub-paragraph (*b*) above if further ring substituted by halogeno, methyl or methoxy groups, but not by other groups;
	(*d*) the salts of any of the substances mentioned in sub-paragraphs (*a*) to (*c*) above;
	(*e*) auramine or magenta.
28 Vibration white finger.	(*a*) The use of hand-held chain saws in forestry; or
	(*b*) the use of hand-held rotary tools in grinding or in the sanding or polishing of metal, or the holding of material being ground, or metal being sanded or polished, by rotary tools; or
	(*c*) the use of hand-held percussive metal-working tools, or the holding of metal being worked upon by percussive tools, in rivetting, caulking, chipping,

Column 1	Column 2
	hammering, fettling or swaging; or
	(d) the use of hand-held powered percussive drills or hand-held powered percussive hammers in mining, quarrying, demolition, or on roads or footpaths, including road construction; or
	(e) the holding of material being worked upon by pounding machines in shoe manufacture.

(This Appendix is reproduced by kind permission of the Controller of HMSO.)

Appendix B

Control of Substances Hazardous to Health Regulations 1988

Regulation 11 Health Surveillance

11.(1) Where it is appropriate for the protection of the health of his employees who are, or are liable to be, exposed to a substance hazardous to health, the employer shall ensure that such employees are under suitable health surveillance.

(2) Health surveillance shall be treated as being appropriate where –

(a) the employee is exposed to one of the substances and is engaged in a process specified in Schedule 5, unless that exposure is not significant; or

(b) the exposure of the employee to a substance hazardous to health is such that an identifiable disease or adverse health effect may be related to the exposure, there is a reasonable likelihood that the disease or effect may occur under the particular conditions of his work and there are valid techniques for detecting indications of the disease or the effect.

(3) The employer shall ensure that a health record, containing particulars approved by the Health and Safety Executive, in respect of each of his employees to whom paragraph (1) relates is made and maintained and that that record or a copy thereof is kept in a suitable form for at least 40 years from the date of the last entry made in it.

(4) Where an employer who holds records in accordance with paragraph (3) ceases to trade, he shall forthwith notify the Health and Safety Executive thereof in writing and offer those records to the Executive.

(5) Subject to regulation 17(4) (which relates to transitional provisions), if an employee is exposed to a substance specified in Schedule 5 and is engaged in a process specified therein, the health surveillance required under paragraph (1) shall include medical surveillance under

the supervision of an employment medical adviser or appointed doctor at intervals of not more than 12 months or at such shorter intervals as the employment medical adviser or appointed doctor may require.

(6) Where an employee is subject to medical surveillance in accordance with paragraph (5) and an employment medical adviser or appointed doctor has certified in the health record of that employee that in his professional opinion that employee should not be engaged in work which exposes him to that substance or that he should only be so engaged under conditions specified in the record, the employer shall not permit the employee to be engaged in such work except in accordance with the conditions, if any, specified in the health record, unless that entry has been cancelled by an employment medical adviser or appointed doctor.

(7) Where an employee is subject to medical surveillance in accordance with paragraph (5) and an employment medical adviser or appointed doctor has certified by an entry in his health record that medical surveillance should be continued after his exposure to that substance has ceased, the employer shall ensure that the medical surveillance of that employee is continued in accordance with that entry while he is employed by the employer, unless that entry has been cancelled by an employment medical adviser or appointed doctor.

(8) On reasonable notice being given, the employer shall allow any of his employees access to the health record which relates to him.

(9) An employee to whom this regulation applies shall, when required by his employer and at the cost of the employer, present himself during his working hours for such health surveillance procedures as may be required for the purposes of paragraph (1) and, in the case of an employee who is subject to medical surveillance in accordance with paragraph (5), shall furnish the employment medical adviser or appointed doctor with such information concerning his health as the employment medical adviser or appointed doctor may reasonably require.

(10) Where, for the purpose of carrying out his functions under these Regulations, an employment medical adviser or appointed doctor requires to inspect any workplace or any record kept for the purposes of these Regulations, the employer shall permit him to do so.

(11) Where an employee or an employer is aggrieved by a decision recorded in the health record by an employment medical adviser or appointed doctor to suspend an employee from work which exposes him to a substance hazardous to health (or to impose conditions on such work), he may, by an application in writing to the Executive within 28 days of the date on which he was notified of the decision, apply for that decision to be reviewed in accordance with a procedure approved for the purposes of this paragraph by the Health and Safety Commission, and the result of that review shall be notified to the employee and

employer and entered in the health record in accordance with the approved procedure.

(12) In this regulation

'appointed doctor' means a fully registered medical practitioner who is appointed for the time being in writing by the Health and Safety Executive for the purposes of this regulation;

'employment medical adviser' means an employment medical adviser appointed under section 56 of the 1974 Act;

"health surveillance" includes biological monitoring.

Control of Substances Hazardous to Health. Approved Code of Practice 1988

Purpose of health surveillance

77   The objectives of health surveillance, where employees are exposed to substances hazardous to health in the course of their work, are:

(a)   the protection of the health of individual employees by the detection at as early a stage as possible of adverse changes which may be attributed to exposure to substances hazardous to health;

(b)   to assist in the evaluation of the measures taken to control exposure;

(c)   the collection, maintenance and use of data for the detection and evaluation of hazards to health;

(d)   to assess, in relation to specific work activities involving micro-organisms hazardous to health, the immunological status of employees.

The results of any health surveillance procedures should lead to some action which will be of benefit to the health of employees. The options and criteria for action should be established before undertaking health surveillance as well as the method of recording, analysis and interpretation of the results of health surveillance.

Suitable health surveillance

78   Health surveillance will always include the keeping of an individual health record (see paragraph 91) and, in addition, it can include a range of procedures, one or more of which is capable of achieving the objectives set out in paragraph 77 above. The procedure(s) which are most suitable in the particular case should be selected. The range

of health surveillance procedures for the purpose of this regulation can be considered to include:

(a)   biological monitoring, i.e. the measurement and assessment of workplace agents or their metabolites either in tissues, secreta, excreta or expired air, or any combination of these in exposed workers;

(b)   biological effect monitoring, i.e. the measurement and assessment of early biological effects in exposed workers;

(c)   medical surveillance, (i.e. both surveillance under the supervision of an employment medical adviser or an appointed doctor for the purpose of regulation 11(5) and under the supervision of a registered medical practitioner) which may include clinical examinations and measurements of physiological and psychological effects of exposure to hazardous substances in the workplace as indicated by alterations in body function or constituents;

(d)   enquiries about symptoms, inspection or examination by a suitably qualified person (e.g. an occupational health nurse);

(e)   inspection by a responsible person (e.g. for chrome ulceration by supervisor, manager, etc);

(f)   review of records and occupational history during and after exposure; the review should be used to check the correctness of the assessment of risks to health made under regulation 6 and indicate whether the assessment requires review.

These procedures are not mutually exclusive and the results of one might indicate the need for another, e.g. the results of biological monitoring may show a need for other health surveillance procedures.

79   Where a method of health surveillance is specified for a particular substance in any Approved Code of Practice under these Regulations, that method should preferably be used.

80   Regulation 11(5) specifies the frequency of medical surveillance carried out under the supervision of employment medical advisers or appointed doctors. This is at intervals not exceeding 12 months, or at such shorter intervals as the employment medical adviser or appointed doctor requires. The exact nature of the examination is at the direction and discretion of the employment medical adviser or appointed doctor.

81   Other health surveillance procedures should be carried out either under the supervision of a registered medical practitioner or, where appropriate, by a suitably qualified person (e.g. an occupational health nurse) or a responsible person. A responsible person is someone appointed by the employer who is competent, in accordance with regulation 12(3), to carry out the relevant procedure and who is charged with reporting to the employer the conclusions of the procedure.

Where health surveillance is appropriate

82   Health surveillance, which must include medical surveillance under the supervision of an employment medical adviser or appointed doctor, is appropriate for workers liable to be exposed to the substances and engaged in the processes listed in Schedule 5 to the regulations.

83   Health surveillance, including the keeping of health records, will also be appropriate for workers exposed to any other substance which fulfils the criteria listed in regulation 11(2)(b). Any judgement as to the likelihood that a disease or adverse health effect may occur must be related to the nature and degree of exposure. The judgement should include assessment of available epidemiology, information on human exposure, and human and animal toxicological data, as well as extrapolation from information about analogous substances or situations.

84   Valid techniques are those of acceptably high sensitivity and specificity which can detect abnormalities related to the nature and degree of exposure. The criteria for interpreting the data should be known (e.g. this may require the establishment of normal values and action levels). The aim should be to establish health surveillance procedures which are safe, easy to perform, non-invasive and acceptable to employees.

85   In the particular conditions of work, should any of the criteria in paragraphs 83 and 84 above not apply, health surveillance procedures should be reviewed and subsequently modified or discontinued as appropriate.

86   Categories where health surveillance is appropriate under the criteria in regulation 11(2)(b) are given below together with information on typical forms of surveillance. Other examples are given in relevant technical literature including HSE Guidance Notes. In all these cases surveillance should be carried out, unless there is no significant risk to health. The list is not definitive and there will be other instances where the criteria in regulation 11(2)(b) indicate that health surveillance is required.

Substance/process	*Typical procedure*
(a) Substances of recognised systemic toxicity.	Appropriate clinical or laboratory investigations.
(b) Substances known to cause occupational asthma	Enquiries seeking evidence of respiratory symptons related to work.
(c) Substances known to cause severe dermatitis	Skin inspection by a responsible person.

(d) (i) Electrolytic plating or
          oxidation of metal articles
          by use of an electrolyte
          containing chromic acid or
          other chromium compounds;

     (ii) Contact with chrome
          solutions in dyeing
          processes using
          dichromate of potassium   } Skin inspection by a responsible person.
          or sodium;

     (iii) Contact with chrome
          solutions in processes of
          liming and tanning of raw
          hides and skins (including
          re-tanning of tanned hides
          or skins).

87 The collection, maintenance and review of health records (see Appendix, para 1(a)) may protect the health of workers through the detection and evaluation of risks to health. In some cases, the only health surveillance required is the collection and maintenance of those records – examples are:

(a) known or suspected carcinogens (e.g. substances listed in Appendix 1 to the Approved Code of Practice for the Control of Carcinogenic Substances and other substances labelled 'may cause cancer') other than those already included in Schedule 5 to the regulations or in paragraph 86 above;

(b) man-made mineral fibres;

(c) dust and fume given off by processes in the rubber industry (other than those included in item (d) in Schedule 5);

(d) leather dust in boot and shoe manufacture, arising during preparation and finishing.

Significant exposure

88 If, following a suitable and sufficient assessment, it can be shown under the circumstances of exposure to a substance hazardous to health, that such exposure is most unlikely to result in any disease or adverse health effect, then exposure can be deemed not to be significant. Further information about significant exposure can be found in other Approved Codes of Practice under these Regulations and in relevant technical literature, including HSE Guidance Notes.

Continuing health surveillance after cessation of exposure

89 In certain circumstances it may be appropriate for an employer to continue health surveillance of his employees (while they remain his employees) after exposure to a substance hazardous to health has ceased. Cases where this will be of benefit to workers may be those where an adverse effect on health may be anticipated after a latent period and where it is believed that the effect can be reliably detected at a sufficiently early stage. Examples of substances which normally

entail continuing health surveillance after cessation of exposure are those which cause cancer of the urinary tract.

Facilities for health surveillance

90 Where health surveillance procedures are carried out at the employer's premises suitable facilities should be available. In cases where examinations and inspections are required, facilities should include a room which is clean, warm, well ventilated, suitably furnished and having a wash basin, equipped with hot and cold running water, soap and clean towel. (If it is not reasonably practicable to provide hot and cold running water, means of heating water should be provided in the room.) The room should be set aside for the exclusive purpose of health surveillance when required and provision should be made for privacy. Where the number of employees to be examined or assessed is substantial, a suitable waiting area should be provided. An adjacent WC with hand-washing facilities should be available for employees when providing specimens for biological monitoring or biological effect monitoring.

Health record

91 A health record, to be kept in all cases where health surveillance is required by the Regulations, should contain at least the information set out in the Appendix to this Code of Practice. These particulars are approved by the Health and Safety Executive.

In a suitable form

92 In addition to keeping the particulars given in the Appendix an index or list of the names of persons undergoing, or who have undergone health surveillance should be kept. The record should be kept in a form compatible with and capable of being linked to those required by regulation 10 for monitoring of exposure, so that, where appropriate, the nature and degree of exposure can be compared with effects on health.

Appendix Health records

Particulars approved by the Health and Safety Executive.

1 A record containing the following particulars should be kept for every employee undergoing health surveillance:

(a) surname, forenames, sex, date of birth, permanent address, post code, National Insurance Number, date of commencement of present employment and a historical record of jobs involving exposure to substances requiring health surveillance in this employment.

(b)   conclusions of all other health surveillance procedures and the date on which and by whom they were carried out. The conclusions should be expressed in terms of the employee's fitness for his work and will include, where appropriate, a record of the decisions of the employment medical adviser or appointed doctor, or conclusions of the medical practitioner, occupational health nurse or other suitably qualified or responsible person, but not confidential clinical data.

2   Where health surveillance consists only of keeping an individual health record the particulars required are those at 1(a) above.

Control of Substances Hazardous to Health Regulations 1988

Schedule 5 Medical surveillance

Column 1 *Substances for which medical surveillance is appropriate*	Column 2 *Processes*
Vinyl chloride monomer (VCM).	In manufacture, production, reclamation, storage, discharge, transport, use or polymerisation.
Nitro or amino derivatives of phenol and of benzene or its homologues.	In the manufacture of nitro or amino derivatives or phenol and of benzene or its homologues and the making of explosives with the use of any of these substances.
Potassium or sodium chromate or dichromate.	In manufacture.
1-Naphthylamine and its salts. Orthotolidine and its salts. Dianisidine and its salts. Dichlorbenzidine and its salts.	In manufacture, formation or use of these substances.
Auramine. Magenta.	In manufacture.
Carbon disulphide. Disulphur dichloride. Benzene, including benzol. Carbon tetrachloride. Trichlorethylene.	Processes in which these substances are used, or given off as vapour, in the manufacture of indiarubber or of articles or goods made wholly or partially of indiarubber.
Pitch.	In manufacture of blocks of fuel consisting of coal, coal dust, coke or slurry with pitch as a binding substance.

(This Appendix is reproduced by kind permission of the Controller of HMSO.)

Appendix C

Prescribed Occupational Diseases

Social Security (Industrial Injuries) (Prescribed Diseases) Regulations 1985, as amended

Prescribed disease or injury	Occupation
A. *Conditions due to physical agents*	Any occupation involving:
1. Inflammation, ulceration or malignant disease of the skin or subcutaneous tissues or of the bones, or blood dyscrasia, or cataract, due to electro-magnetic radiations (other than radiant heat), or to ionising particles.	Exposure to electro-magnetic radiations (other than radiant heat) or to ionising particles.
2. Heat cataract.	Frequent or prolonged exposure to rays from molten or red-hot material.
3. Dysbarism, including decompression sickness, barotrauma and osteonecrosis.	Subjection to compressed or rarified air or other respirable gases or gaseous mixtures.
4. Cramp of the hand or forearm due to repetitive movements.	Prolonged periods of handwriting, typing or other repetitive movements of the fingers, hand or arm.
5. Subcutaneous cellulitis of the hand (Beat hand).	Manual labour causing severe or prolonged friction or pressure on the hand.
6. Bursitis or subcutaneous cellulitis arising at or about the knee due to severe or prolonged external friction or pressure at or about the knee (Beat knee).	Manual labour causing severe or prolonged external friction or pressure at or about the knee.
7. Bursitis or subcutaneous cellulitis arising at or about the elbow due to severe or prolonged external friction or pressure at or about the elbow (Beat elbow).	Manual labour causing severe or prolonged external friction or pressure at or about the elbow.
8. Traumatic inflammation of the tendons of the hand or forearm, or of the associated tendon sheaths.	Manual labour, or frequent or repeated movements of the hand or wrist.
9. Miner's nystagmus.	Work in or about a mine.

313

Prescribed disease or injury	Occupation
10. Substantial sensorineural hearing loss amounting to at least 50dB in each ear, being due in the case of at least one ear to occupational noise, and being the average of pure tone loss measured by audiometry over the 1, 2 and 3 kHz frequencies (occupational deafness).	(a) The use of, or work wholly or mainly in the immediate vicinity of, pneumatic percussive tools or high-speed grinding tools, in the cleaning, dressing or finishing of cast *metal* or of ingots, billets or blooms (but not stone/concrete used in road/railway construction or

Any occupation involving:

(b) the use of, or work wholly or mainly in the immediate vicinity of, pneumatic percussive tools on metal in the shipbuilding or ship repairing industries; or

(c) the use of, or work in the immediate vicinity of, pneumatic percussive tools on metal, or for drilling rock in quarries or underground, or in mining coal, for at least an average of one hour per working day; or

(d) work wholly or mainly in the immediate vicinity of drop-forging plant (including plant for drop-stamping or drop-hammering) or forging press plant engaged in the shaping of metal; or

(e) work wholly or mainly in rooms or sheds where there are machines engaged in weaving man-made or natural (including mineral) fibres or in the bulking up of fibres in textile manufacturing; or

(f) the use of, or work wholly or mainly in the immediate vicinity of, machines engaged in cutting, shaping or cleaning metal nails; or

(g) the use of, or work wholly or mainly in the immediate vicinity of, plasma spray guns engaged in the deposition of metal; or

(h) the use of, or work wholly or mainly in the immediate vicinity of, any of the following machines engaged in the working of wood or material composed partly of wood, that is to say; multi-cutter moulding machines, planning machines, automatic or semi-automatic lathes, multiple cross-cut machines, automatic shaping machines, double-end tenoning machines, vertical spindle moulding machines (including high-speed routing machines), edge banding machines, bandsawing machines with

Prescribed disease or injury	Occupation
	a blade width of not less than 73 millimetres and circular sawing machines in the operation of which the blade is moved towards the material being cut; or (*i*) the use of chain saws in forestry. Any occupation involving:
11. Episodic blanching, occurring throughout the year, affecting the middle or proximal phalanges or in the case of a thumb the proximal phalanx, of – (*a*) in the case of a person with 5 fingers (including thumb) on one hand, any 3 of those fingers, or (*b*) in the case of a person with only 4 such fingers, any 2 of those fingers, or (*c*) in the case of a person with less than 4 such fingers, any one of those fingers or, as the case may be, the one remaining finger (vibration white finger).	(*a*) The use of hand-held chain saws in forestry; or (*b*) the use of hand-held rotary tools in grinding or in the sanding or polishing of metal, or the holding of material being ground, or metal being sanded or polished, by rotary tools; or (*c*) the use of percussive metal-working tools, or the holding of metal being worked upon by percussive tools, in riveting, caulking, chipping hammering, fettling or swaging; or (*d*) the use of hand-held powered percussive drills or hand-held powered percussive hammers in mining, quarrying, demolition, or on roads or footpaths, including road construction; or (*e*) the holding of material being worked upon by pounding machines in shoe manufacture.
12. Carpal tunnel syndrome.	The use of hand-held vibrating tools.
B. *Conditions due to biological agents*	
1. Anthrax.	Contact with animals infected with anthrax or the handling (including the loading or unloading or transport) of animal products or residues.
2. Glanders.	Contact with equine animals or their carcases.
3. Infection by leptospira.	(*a*) Work in places which are, or are liable to be, infested by rats, field mice or voles, or other small mammals; or (*b*) work at dog kennels or the care or handling of dogs; or (*c*) contact with bovine animals or their meat products or pigs or their meat products.
4. Ankylostomiasis.	Work in or about a mine.
5. Tuberculosis.	Contact with a source of tuberculosis infection.

Prescribed disease or injury	Occupation
	Any occupation involving:
6. Extrinsic allergic alveolitis (including farmer's lung).	Exposure to moulds or fungal spores or heterologous proteins by reason of employment in:
	(*a*) agriculture, horticulture, forestry, cultivation of edible fungi or malt-working; or
	(*b*) loading or unloading or handling in storage mouldy vegetable matter or edible fungi; or
	(*c*) caring for or handling birds; or
	(*d*) handling bagasse.
7. Infection by organisms of the genus brucella.	Contact with –
	(*a*) animals infected by brucella, or their carcases or parts thereof, or their untreated products; or
	(*b*) laboratory specimens or vaccines of, or containing, brucella.
8. Viral hepatitis.	Close and frequent contact with –
	(*a*) human blood or human blood products; or
	(*b*) a source of viral hepatitis.
9. Infection by *Streptococcus suis*	Contact with pigs infected by *Streptococcus suis*, or with the carcases, products or residues of pigs so infected.
10. (*a*) Avian chlamydiosis	Contact with birds infected with chlamydia psittaci, or with the remains or untreated products of such birds.
(*b*) Ovine chlamydiosis	Contact with sheep infected with chlamydia psittaci, or with the remains or untreated products of such sheep.
11. Q fever	Contact with animals, their remains or their untreated products.
12. Orf	Contact with sheep, goats or with the carcases of sheep or goats.
13. Hydatidosis	Contact with dogs.
C. *Conditions due to chemical agents*	
1. Poisoning by lead or a compound of lead.	The use or handling of, or exposure to the fumes, dust or vapour of, lead or a compound of lead, or a substance containing lead.
	Any occupation involving:
2. Poisoning by manganese or a compound of manganese.	The use of handling of, or exposure to the fumes, dust or vapour of, manganese or a compound of manganese, or a substance containing manganese.

Prescribed disease or injury	Occupation
3. Poisoning by phosphorus or an inorganic compound or phosphorus or poisoning due to the anticholinesterase or pseudo anticholinesterase action of organic phosphorus compounds.	The use or handling of, or exposure to the fumes, dust or vapour of, phosphorus or a compound of phosphorus, or a substance containing phosphorus.
4. Poisoning by arsenic or a compound of arsenic.	The use or handling of, or exposure to the fumes, dust or vapour of, arsenic or a compound of arsenic, or a substance containing arsenic.
5. Poisoning by mercury or a compound of mercury.	The use of handling of, or exposure to the fumes, dust or vapour of, mercury or a compound of mercury, or a substance containing mercury.
6. Poisoning by carbon bisulphide.	The use or handling of, or exposure to the fumes or vapour of, carbon bisulphide or a compound of carbon bisulphide, or a substance containing carbon bisulphide.
7. Poisoning by benzene or a homologue of benzene.	The use or handling of, or exposure to the fumes of, or vapour containing benzene or any of its homologues.
8. Poisoning by nitro- or amino- or chloro-derivative of benzene or of a homologue of benzene, or poisoning by nitrochlorbenzene.	The use or handling of, or exposure to the fumes of, or vapour containing a nitro-or amino- or chloro- derivative of benzene, or of a homologue of benzene, or nitrochlorbenzene.
9. Poisoning by dinotrophenol or a homologue of dinitrophenol or by substituted dinitrophenols or by the salts of such substances.	The use or handling of, or exposure to the fumes of, or vapour containing, dinitrophenol or a homologue or substituted dinitrophenols or the salts of such substances.
10. Poisoning by tetrachloroethane.	The use or handling of, or exposure to the fumes of, or vapour containing, tetrachloroethane.
11. Poisoning by diethylene dioxide (dioxan).	The use or handling of, or exposure to the fumes of, or vapour containing, diethylene dioxide (dioxan).
12. Poisoning by methyl bromide.	The use or handling of, or exposure to the fumes of, or vapour containing, methyl bromide.
	Any occupation involving:
13. Poisoning by chlorinated naphthalene.	The use or handling of, or exposure to the fumes of, or dust or vapour containing, chlorinated nephthalene.
14. Poisoning by nickel carbonyl.	Exposure to nickel carbonyl gas.
15. Poisoning by oxides of nitrogen.	Exposure to oxides of nitrogen.

Prescribed disease or injury	Occupation
16. Poisoning by gonioma kamassi (African boxwood).	The manipulation of gonioma kamassi or any process in or incidental to the manufacture of articles therefrom.
17. Poisoning by beryllium or a compound of beryllium.	The use or handling of, or exposure to the fumes of, or dust or vapour of, beryllium or a compound of beryllium, or a substance containing beryllium.
18. Poisoning by cadmium.	Exposure to cadmium dust or fumes.
19. Poisoning by acrylamide monomer.	The use or handling of, or exposure to, acrylamide monomer.
20. Dystrophy of the cornea (including ulceration of the corneal surface) of the eye.	(*a*) The use or handling of, or exposure to, arsenic, tar, pitch, bitumen, mineral oil (including paraffin), soot or any compound, product or residue of any of these substances, except quinone or hydroquinone; or (*b*) exposure to quinone or hydroquinone during their manufacture.
21. (*a*) Localised new growth of the skin, papillomatous or keratotic; (*b*) squamous-celled carcinoma of the skin.	The use or handling of, or exposure to, arsenic, tar, pitch, bitumen, mineral oil (including paraffin), soot or any compound, product or residue of any of these substances, except quinone or hydroquinone.
22. (*a*) Carcinoma of the mucous membrane of the nose or associated air sinuses; (*b*) primary carcinoma of a bronchus or of a lung.	Work in a factory where nickel is produced by decomposition of a gaseous nickel compound which necessitates working in or about a building or buildings where that process or any other industrial process ancillary or incidental thereto is carried on.

23. Primary neoplasm (including papilloma, carcinoma-in-situ and invasive carcinoma) of the epithelial lining of the urinary tract (renal pelvis, ureter, bladder and urethra).	(*a*) Work in a building in which any of the following substances is produced for commercial purposes:
	  (i) alpha-naphthylamine, beta-naphthylamine or methyl-enebis-orthochloroaniline;
	  (ii) diphenyl substituted by at least one nitro or primary amino group or by at least one nitro and primary amino group (including benzidine);
	  (iii) any of the substances mentioned in sub-paragraph (ii) above if further ring substituted by halogeno, methyl or methoxy groups, but not by other groups;
	  (iv) the salts of any of the substances mentioned in sub-paragraphs (i) to (iii) above;
	  (v) auramine or magenta; or

Prescribed disease or injury	Occupation
	Any occupation involving:
	(b) the use or handling of any of the substances mentioned in sub-paragraph (a) (i) to (iv), or work in a process in which any such substance is used, handled or liberated; or
	(c) the maintenance or cleaning of any plant or machinery used in any such process as is mentioned in sub-paragraph (b), or the cleaning of clothing used in any such buildings as is mentioned in sub-paragraph (a) if such clothing is cleaned within the works of which the building forms a part or in a laundry maintained and used solely in connection with such works.
24. (a) Angiosarcoma of the liver; (b) osteolysis of the terminal phalanges of the fingers; (c) non-cirrhotic portal fibrosis.	(a) Work in or about machinery or apparatus used for the polymerization of vinyl chloride monomer, a process which, for the purposes of this provision, comprises all operations up to and including the drying of the slurry produced by the polymerization and the packaging of the dried product; or (b) work in a building or structure in which any part of that process takes place.
25. Occupational vitiligo.	The use or handling of, or exposure to, para-tertiary-butylphenol, para-tertiary-butylcatechol, para-amyl-phenol, hydroquinone or the monobenzyl or monobutyl ether of hydroquinone.
	Any occupation involving:
26. Damage to the liver or kidneys due to exposure to carbon tetrachloride.	Use of or handling of or exposure to the fumes of, or vapour containing carbon tetrachloride.
27. Damage to the liver or kidneys due to exposure to trichloromethane (chloroform).	Use of or handling of or exposure to fumes of or vapour containing trichloromethane (chloroform).
28. Central nervous system dysfunction and associated gastro-intestinal disorders due to exposure to chloromethane (methyl chloride).	Use of or handling of or exposure to fumes or vapours containing chloromethane (methyl chloride).
29. Peripheral neuropathy due to exposure to n-hexane or methyl n-butyl ketone.	Use of or handling of or exposure to the fumes of or vapours containing n-hexane or methyl n-butyl ketone.

Prescribed disease or injury	Occupation

D. *Miscellaneous Conditions*

1. Pneumoconiosis.

Any occupation –

(*a*) set out in Part II of Schedule 1 of the 1980 Regulations;
(*b*) specified in regulation 2(*b*)(ii); (i.e. relating to processes involving exposure to dust).

Any occupation involving:

2. Byssinosis.

Work in any room where any process up to and including the weaving process is performed in a factory in which the spinning or manipulation of raw or waste cotton or of flax, or the weaving of cotton or flax, is carried on.

3. Diffuse mesothelioma (primary neoplasm of the mesothelium of the pleura or of the pericardium or of the peritoneum).

(*a*) The working or handling of asbestos or any admixture of asbestos; or
(*b*) the manufacture or repair of asbestos textiles or other articles containing or composed of asbestos; or
(*c*) the cleaning of any machinery or plant used in any of the foregoing operations and of any chambers, fixtures and appliances for the collection of asbestos dust; or
(*d*) substantial exposure to the dust arising from any of the foregoing operations.

4. Inflammation or ulceration of the mucous membranes of the upper respiratory passages or mouth produced by dust, liquid or vapour.

Exposure to dust, liquid or vapour.

Any occupation involving:

5. Non-infective dermatitis of external origin (including chrome ulceration of the skin but excluding dermatitis due to ionising particles or electro-magnetic radiations other than radiant heat).

Exposure to dust, liquid or vapour or any other external agent capable of irritating the skin (including friction or heat but excluding ionising particles or electro-magnetic radiations other than radiant heat).

6. Carcinoma of the nasal cavity or associated air sinuses (nasal carcinoma).

(*a*) Attendance for work in, on or about a building where wooden goods are manufactured or repaired; or
(*b*) attendance for work in a building used for the manufacture of footwear or components of footwear made wholly or partly of leather or fibre board; or
(*c*) attendance for work at a place used wholly or mainly for the repair of footwear made wholly or partly of leather or fibre board.

Prescribed disease or injury	Occupation
7. Asthma which is due to exposure to any of the following agents:	Exposure to any of the agents set out in column 1 of this paragraph.

    (*a*) isocyanates;

    (*b*) platinum salts;

    (*c*) fumes or dusts arising from the manufacture, transport or use of hardening agents (including epoxy resin curing agents) based on phthalic anhydride, tetrachlorophthalic anhydride, trimellitic anhydride or triethylene-tetramine;

    (*d*) fumes arising from the use of rosin as a soldering flux;

    (*e*) proteolytic enzymes;

    (*f*) animals including insects and other anthropods used for the purposes of research or education or in laboratories;

    (*g*) dusts arising from the sowing, cultivation, harvesting, drying, handling, milling, transport or storage of barley, oats, rye, wheat or maize, or the handling, milling, transport or storage of meal or flour made therefrom;

    (*h*) antibiotics;

    (*i*) cimetidine;

    (*j*) wood dust;

    (*k*) ispaghula;

    (*l*) castor bean dust;

    (*m*) ipecacuanha;

    (*n*) azodicarbonamide;

    (*o*) animals including insects and other arthropods or their larval forms, used for the purposes of pest control or fruit cultivation, or the larval forms of animals used for the purposes of research, education or in laboratories;

    (*p*) glutaraldehyde;

    (*q*) persulphate salts or henna;

    (*r*) crustaceans or fish or products arising from these in the food processing industry;

    (*s*) reactive dyes;

    (*t*) soya bean;

    (*u*) tea dust;

    (*v*) green coffee bean dust;

    (*w*) fumes from stainless steel welding;

    (*x*) any other sensitising agent; (occupational asthma).

Prescribed disease or injury	Occupation

Moreover, the time at which a person shall be treated as having developed prescribed diseases B.12 and B.13, or occupational asthma due to exposure to agents specified in D.7(*o*) to (*x*), is the first day on which that person is incapable of work, or suffering from a loss of faculty as a result of those diseases after 25 September 1991.

Any occupation involving:

8. Primary carcinoma of the lung where there is accompanying evidence of one or both of the following:
   (*a*) asbestosis;
   (*b*) bilateral diffuse pleural thickening.

(*a*) The working or handling of asbestos or any admixture of asbestos; or
(*b*) the manufacture or repair of asbestos textiles or other articles containing or composed of asbestos; or
(*c*) the cleaning of any machinery or plant used in any of the foregoing operations and of any chambers, fixtures and appliances for the collection of asbestos dust; or
(*d*) substantial exposure to the dust arising from any of the foregoing operations.

9. Bilateral diffuse pleural thickening.

(*a*) The working or handling of asbestos or any admixture of asbestos; or
(*b*) the manufacture or repair of asbestos textiles or other articles containing or composed of asbestos; or
(*c*) the cleaning of any machinery or plant used in any of the foregoing operations and of any chambers, fixtures and appliances for the collection of asbestos dust; or
(*d*) substantial exposure to the dust arising from any of the foregoing operations.

10. Primary carcinoma of the lung.

(*a*) Work underground in a tin mine; or
(*b*) exposure to bis (chloromethyl) ether produced during the manufacture of chloromethyl methyl ether; or
(*c*) exposure to pure zinc chromate, calcium chromate or strontium chromate.

11. Primary carcinoma of the lung where there is accompanying evidence of silicosis.

Exposure to silica dust in the course of:

(*a*) The manufacture of glass or pottery;
(*b*) tunnelling in or quarrying sandstone or granite;
(*c*) mining metal ores;
(*d*) slate quarrying or the manufacture of artefacts from slate;
(*e*) mining clay;
(*f*) using siliceous materials as abrasives;
(*g*) cutting stone;
(*h*) stonemasonry; or
(*i*) work in a foundry.

Prescribed disease or injury	Occupation
12. (a) Chronic bronchitis; or (b) emphysema; or (c) both.	Exposure to coal dust by reason of working underground in a coal mine for a period of, or periods amounting in the aggregate to, at least 20 years (whether before or after 5 July 1948).

(This Appendix is reproduced by kind permission of the Controller of HMSO.)

Appendix D

Management of Health and Safety at Work Regulations 1992 and Approved Code of Practice

Regulation 5 Health Surveillance

Every employer shall ensure that his employees are provided with such health surveillance as is appropriate having regard to the risks to their health and safety which are identified by the assessment.

Approved Code of Practice

30 The risk assessment will identify circumstances in which health surveillance is required by specific health and safety regulations (e.g. COSHH, Asbestos). In addition, health surveillance should be introduced where the assessment shows the following criteria to apply:

(*a*) there is an identifiable disease or adverse health condition related to the work concerned;

(*b*) valid techniques are available to detect indications of the disease or condition;

(*c*) there is a reasonable likelihood that the disease or condition may occur under the particular conditions of work; and

(*d*) surveillance is likely to further the protection of the health of the employees to be covered.

31 The primary benefit, and therefore the objective, of health surveillance should be to detect adverse health effects at an early stage, thereby enabling further harm to be prevented. In addition the results of health surveillance can provide a means of:

(*a*) checking the effectiveness of control measures;

(*b*) providing feedback on the accuracy of the risk assessment;

(*c*) identifying and protecting individuals at increased risk.

32 Once it is decided that health surveillance is appropriate, such health surveillance should be maintained during the employee's employment unless the risk to which the worker is exposed and associated health effects are short term. The minimum requirement for health surveillance is the keeping of an individual health record. Where it is appropriate,

324

health surveillance may also involve one or more health surveillance procedures depending on their suitability in the circumstances. Such procedures can include:

(*a*) inspection of readily detectable conditions by a responsible person acting within the limits of their training and experience;

(*b*) enquiries about symptoms, inspection and examination by a qualified person such as an Occupational Health Nurse;

(*c*) medical surveillance, which may include clinical examination and measurements of physiological or psychological effects by an appropriately qualified practitioner;

(*d*) biological effect monitoring, ie the measurement and assessment of early biological effects such as diminished lung function in exposed workers;

(*e*) biological monitoring, i.e. the measurement and assessment of workplace agents or their metabolites either in tissues, secreta, excreta, expired air or any combination of these in exposed workers.

33 The frequency of the use of such methods should be determined either on the basis of suitable general guidance (e.g. as regards skin inspection for dermal effects) or on the advice of a qualified practitioner; the employees concerned should be given an opportunity to comment on the proposed frequency of such health surveillance procedures and should have access to an appropriately qualified practitioner for advice on surveillance.

(This Appendix is reproduced by kind permission of the Controller of HMSO.)

Index